Ben Stacy Jerrik (Ed.)

Ronald Brown (Mathematician)

Ben Stacy Jerrik (Ed.)

Ronald Brown (Mathematician)

George Gargrave, Henry Gellibrand, Peter Giblin

Part Press

Publisher:
Part Press is a trademark of
International Book Market Service Ltd., 17 Rue Meldrum, Beau Bassin, 1713-01 Mauritius
Email: info@bookmarketservice.com
Website: www.bookmarketservice.com

Published in 2011

Printed in: U.S.A., U.K., Germany. This book was not produced in Mauritius.

ISBN: 978-613-9-29294-3

Contents

Articles

Ronald_Brown_(mathematician) 1
George_Gargrave 4
William_Gascoigne_(scientist) 4
Henry_Gellibrand 6
Gerland_(mathematician) 7
Peter_Giblin 8
Richard_D._Gill 9
James_Whitbread_Lee_Glaisher 11
Alfred_Goldie 12
Benjamin_Gompertz 13
I._J._Good 16
Reuben_Goodstein 20
John_Gough_(natural_philosopher) 21
Timothy_Gowers 23
Jeremy_Gray 27
Ralph_Greatorex 28
John_Greaves 29
Ben_J._Green 33
George_Green 35
Olinthus_Gregory 41
John_Greig_(mathematician) 42

References

Article Sources and Contributors 43
Image Sources, Licenses and Contributors 44

Ronald_Brown_(mathematician)

Ronald M. Brown	
Born	4 January 1935London
Nationality	United Kingdom
Fields	Mathematics
Alma mater	University of Oxford
Doctoral advisor	J. H. C. Whitehead

Ronald Brown is an English mathematician. Emeritus Professor in the School of Computer Science at Bangor University,[1] he has authored many books and journal articles.

Education and career

Born on 4 January 1935 in London, Brown attended Oxford University, obtaining a B.A. in 1956 and a Ph.D. in 1962.[2] Brown began his teaching career during his Ph.D. work, serving as an assistant lecturer at Liverpool University before assuming the position as lecturer. In 1964, he took a position at Hull University, serving first as a senior lecturer and then as a reader before becoming a professor of pure mathematics at Bangor University, then a part of the University of Wales, in 1970.

Brown served as professor of pure mathematics for 29 years, also serving briefly during the 1983-1984 term as an associate professor at Louis Pasteur University in Strasbourg.[2] In 1999, Brown took a half-time research professorship until he became Professor Emeritus in 2001.

Editing and writing

Brown has served as an editor or on the editorial board for a number of print and electronic journals. He began in 1968 with the *Chapman & Hall Mathematics Series*, contributing through 1986.[2] In 1975, he joined the editorial advisory board of the London Mathematical Society, remaining through 1994. Two years later, he joined the editorial board of Applied Categorical Structures,[3] continuing through 2007. From 1995 and 1999, respectively, he has been active with the electronic journals [*Theory and Applications of Categories* http://www.tac.mta.ca/tac/][4] and *[Homology, Homotopy and Applications* http://www.intlpress.com/ *]*,[5] which he helped found. Since 2006, he has been involved with *Journal of Homotopy and Related Structures*.[6] [7] His mathematical research interests range from algebraic topology and groupoids, to homology theory, category theory, mathematical biology, mathematical physics and higher dimensional algebra.[8] [9] [10] [11] [12]

Brown has authored or edited a number of books and over 150 academic papers published in academic journals or collections. His first published paper was "Ten topologies for $X\times Y$", which was published in the *Quarterly Journal of Math* in 1963[13] Since then, his publications have appeared in many journals, including but not limited to the *Journal of Algebra*, *Proceedings of the American Mathematical Society*, *Mathematische Zeitschrift*, *College Mathematics Journal*, and *American Mathematical Monthly*. He is also known for several recent co-authored papers on Categorical ontology.[14]

Among his several books and standard topology and algebraic topology textbooks are: *Elements of Modern Topology* (1968), *Low-Dimensional Topology* (1979, co-edited with T.L. Thickstun), *Topology: a geometric account of general topology, homotopy types, and the fundamental groupoid* (1998),[15] [16] *Topology and Groupoids* (2006)[17] and *Nonabelian Algebraic Topology: Filtered Spaces, Crossed Complexes, Cubical Homotopy Groupoids* (EMS, 2010).[17] [18] [19] [20] [21] [22] [23] [24] [25] [26]

His recent fundamental results that extend the classical Van Kampen theorem to higher homotopy in higher dimensions (HHSvKT) are of substantial interest for solving several problems in algebraic topology, both old and new.[27] Moreover, developments in algebraic topology have often had wider implications,as for example in algebraic geometry and also in algebraic number theory. Such higher dimensional (HHSvKT) theorems are about homotopy invariants of structured spaces, and especially those for filtered spaces or n-cubes of spaces. A nice example is the fact that the relative Hurewicz theorem is a consequence of HHSvKT, and this then suggested a triadic Hurewicz theorem.

See also

- Higher dimensional algebra
- Higher category theory
- Seifert-van Kampen theorem
- Groupoids
- Algebraic topology
- Nonabelian algebraic topology
- R-algebroids
- Double groupoids
- Homology (mathematics)
- Alexander Grothendieck
- arXiv

References

[1] http://www.bangor.ac.uk/~mas010/Page at Bangor University, UK

[2] "Ronald Brown: Biographical Sketch" (http://www.bangor.ac.uk/~mas010/biog-br.html). *Bangor University*. . Retrieved 2010-04-23.

[3] *"Applied Categorical Structures"* (http://www.bangor.ac.uk/~mas010/publicfull.htm). *Springer Publishers, Berlin*. 2010-12-08. . Retrieved 2010-12-11.

[4] EDITORIAL BOARD of the International journal *Theory and Applications of Categories* http://www.tac.mta.ca/tac/geninfo.html#edlist

[5] EDITORIAL BOARD of the International journal *Homology, Homotopy and Applications* http://www.intlpress.com/HHA/editors.htm

[6] http://arxiv.org/

[7] Listed in the Editors' List of the International *Journal of Homotopy and Related Structures* http://tcms.org.ge/Journals/JHRS/editors.htm

[8] Editor Ronald Brown's research interests: Category theory, higher dimensional algebra, holonomy, groupoids and crossed objects in algebraic topology. http://emis.kaist.ac.kr/journals/JHRS/interests.htm

[9] Cited by John C. Baez, James Dolan., in Higher-Dimensional Algebra III: n-categories and the Algebra of Opetopes.,quantum algebra and Topology, Adv. Math. 135 (1998), 145-206.

[10] Cited by Georgescu, George and Popescu, Andrei., in ``A common generalization for MV-algebras and Łukasiewicz-Moisil algebras", *Archive for Mathematical Logic*, Vol. 45, No. 8. (November 2006), pp. 947-981. (in reference to Heyting-algebra higher-dimensional-algebra hyperalgebras Łukasiewicz-Moisil-algebras meta-logics MV-algebras on 2007-07-11)

[11] Cited by John C. Baez, Laurel Langford., in ``Higher-Dimensional Algebra IV: 2-Tangles.",(*Quantum Algebra* (math.QA); *Algebraic Topology* (math.AT); *Category Theory* (math.CT)), **Adv. Math**. 180 (2003), 705-764 http://www.azimuthproject.org/azimuth/show/John+Baez

[12] Cited in ``*Higher-dimensional Algebra and Topological Quantum Field Theory*" **J.Math.Phys**. **36** (1995) 6073-6105, by John C. Baez, James Dolan, (2004) http://arxiv.org/abs/q-alg/9503002 doi:10.1063/1.531236

[13] "Ronald Brown Publications" (http://web.archive.org/web/20080424002106/http://www.bangor.ac.uk/~mas010/publicfull.htm). *Bangor University*. 2008-04-19. Archived from the original (http://www.bangor.ac.uk/~mas010/publicfull.htm) on 2008-04-24. . Retrieved 2010-04-23.

[14] Cited in Online research in philosophy Entries: http://philpapers.org/

[15] Cited in Encyclopaedia of Mathematics - ISBN 1402006098

[16] Referenced in ``**Algebraic homotopy**" http://eom.springer.de/A/a130170.htm

[17] Cited in ``**Bibliography For Groupoids And Algebraic Topology**" http://myyn.org/m/article/bibliography-for-groupoids-and-algebraic-topology/

[18] http://sz0009.ev.mail.comcast.net/service/home/~/Tracts_vol15.pdf?auth=co&loc=en_US&id=128480&part=2 and www.ems-ph.org : *EMS Tracts in Mathematics*, Vol. 15

[19] http://citeseerx.ist.psu.edu/viewdoc/summary?doi=10.1.1.150.6444 CiteSeerX accessed on 12/10/2010
[20] http://ncatlab.org/nlab/show/Nonabelian+Algebraic+Topology nLab Review of Nonabelian Algebraic Topology
[21] A Review of ``*Nonabelian Algebraic Topology*" by Prof. J. Baez on June 6, 2009 http://golem.ph.utexas.edu/category/2009/06/nonabelian_algebraic_topology.html
[22] Addebook • Apr 22nd, 2009 • Category: Mathematics : *Nonabelian Algebraic Topology* http://www.addebook.com/tech2/mathematics/nonabelian-algebraic-topology_4164.html
[23] Cited on p. xi as a basic reference in ``*Non-abelian Theories*" http://myyn.org/m/article/non-abelian-theories/
[24] [ALGTOP-L] available full draft of book on `**Nonabelian algebraic topology**' https://lists.lehigh.edu/pipermail/algtop-l/2009q2/000443.html
[25] Cited in ``Towards Higher Categories" By John C. Baez and J. Peter May, Publisher: Springer Verlag,Published Date: 2009-10-01, ISBN 9781441915238 http://www.isbnlib.com/author/John_C__Baez
[26] Referenced in ``**Nonabelian cohomology**" in nLab, Revised on November 9, 2010 00:20:06 by Urs Schreiber http://ncatlab.org/nlab/show/nonabelian+cohomology
[27] The higher Van Kampen Theorems and computation of the unstable homotopy groups of spheres and complex spaces http://mathoverflow.net/questions/39818/the-higher-van-kampen-theorems-and-computation-of-the-unstable-homotopy-groups-of

External links

- Ronald Brown's Biography and publications (http://planetphysics.org/encyclopedia/RonaldBrown.html)
- Ronald Brown's Home Page (http://www.bangor.ac.uk/~mas010/)
- Higher-Dimensional Algebra citations list (http://www.citeulike.org/tag/higher-dimensional-algebra)
- Mathematical Genealogy Project page (http://genealogy.math.ndsu.nodak.edu/id.php?id=45042)
- MathOverflow-- A place for mathematicians to ask and answer questions. (http://mathoverflow.net/)
- *Editorial Board of Journal of Homotopy and Related Structures (JHRS)* (http://tcms.org.ge/Journals/JHRS/editors.htm)
- nLab Abstract Mathematics Website (http://ncatlab.org/nlab/)
- Editorial Board of *Homology, Homotopy and Applications (HHA)* (http://www.intlpress.com/HHA/)
- The Origins of `Pursuing Stacks' by Alexander Grothendieck (http://www.bangor.ac.uk/~mas010/pstacks.htm)
- Journal of Homotopy and Related Structur(JHRS) Publications (http://arxiv.org/)
- Homology, Homotopy and Applications (http://www.intlpress.com/)
- Theory and Applications of Categories (http://www.tac.mta.ca/tac/)

George_Gargrave

George Gargrave (1710–1785), was an English mathematician.

Gargrave, born at Leyburn, Yorkshire, in 1710, was educated by his uncle, John Crow, a schoolmaster in that place. Under him he acquired a considerable knowledge of the classics and mathematics. His natural bent was towards astronomy, and in after life he was reputed one of the best proficients in the less recondite branches of that science in the north of England. In 1745 he became associated with Joseph Randall in the management of the academy at Heath, near Wakefield. The academy, though of good repute, did not pay, and was given up in 1754. Gargrave then started at Wakefield a mathematical school, with such success that in 1768 he retired on a handsome competency. He died on 7 December 1785, and was buried in the churchyard at Wensley. Gargrave was a musician of some skill, and his handwriting was remarkably clear and fine. He possessed a large and well-selected library, and a fine collection of astronomical and other scientific apparatus. He contributed to the *Gentleman's Magazine* a translation of Dr. Halley's 'Dissertation on the Transit of Venus' (1760, p. 265); 'Observations on the Transit of Venus' (1761, p. 296); on the same subject (1769, pp. 278–9); 'Observations of an Eclipse of the Moon' (1776, p. 357); and 'Memoirs of Mr. Abraham Sharp, mathematician, mechanic, and astronomer' (1781, p. 461). He also left a manuscript treatise on the doctrine of the sphere.

References

This article incorporates text from a publication now in the public domain: "Gargrave, George". *Dictionary of National Biography*. London: Smith, Elder & Co. 1885–1900.

William_Gascoigne_(scientist)

William Gascoigne (1612 – 2 July 1644) was an English astronomer, mathematician and maker of scientific instruments from Middleton near Leeds who invented the micrometer. He was one of "nos Keplari" a group of astronomers in the north of England who followed the astronomy of Johannes Kepler which included, Jeremiah Horrocks and William Crabtree.

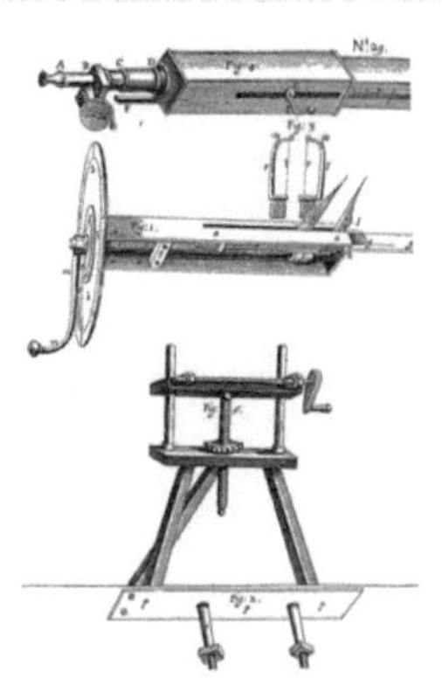

Gascoigne's Micrometer as drawn by Robert Hooke

Life and work

Gascoigne was born in Middleton, Leeds in 1612, the son of a minor country gentleman. His father was Henry Gascoigne, Esq., of Thorpe-on-the-Hill in the parish of Rothwell, near Leeds, Yorkshire. His mother was Margaret Jane, daughter of William Cartwright. Little is known of his early life. He claimed he was educated at the University of Oxford, although no record of this has been found.[1]

In the late 1630s, Gascoigne, was working on a Keplerian optical arrangement when a thread from a spider's web happened to become caught at exactly the combined optical focal points of the two lenses. When he looked through the arrangement Gascoigne saw the web bright and sharp within the field of view. He realized that he could more accurately point the telescope using the line as a guide, and went on to invent the telescopic sight by placing crossed wires at the focal point to define the centre of the field of view.[2] He then added this arrangement to a sextant modelled on the instrument used by Tycho Brahe, although Tycho's sextant was only a naked-eye instrument. Gascoigne's sextant was five feet in radius, and measured the distance between astronomical bodies to an unprecedented degree of accuracy. Gascoigne then realised that by introducing two points, whose separation could be adjusted using a screw, he could measure the size of the image enclosed by them. Using the known pitch of the screw, and knowing the focal length of the lens producing the image, he could work out the size of the object, such as the Moon or the planets, to a hitherto unattainable degree of accuracy.[1]

Gascoigne met the Lancashire astronomer William Crabtree, probably in 1640. After making observations at Gascoigne's home, Crabtree was much taken with these inventions and immediately saw their significance. On his return to his home in Broughton Spout, just outside Manchester, he wrote to Gascoigne asking if he might obtain such instruments and also wrote to his friend Jeremiah Horrocks about them. He wrote again to Gascoigne on 28 December 1640 saying,

> My friend Mr. Horrox professeth that little touch which I gave him hath ravished his mind quite from itself and left him in an Exstasie between Admiration and Amazement. I beseech you Sir, slack not your Intentions for the Perfection of your begun Wonders.

Sadly Horrocks died before he could try out the instruments, but Crabtree and Gascoigne did use them to try to corroborate Horrocks's theories about the elliptical orbit of the Moon.[1]

This invention was later taken up and improved by the scientist and astronomer Richard Towneley who was the nephew of Gascoigne's friend Christopher Towneley. Towneley later brought the instrument to the attention of Robert Hooke, who used it to calculate the size of comets and other celestial bodies. The micrometer, as it became known, was to lie at the heart of astronomical measurement down to the twentieth century.[3]

In 1642, civil war broke out in England, and Gascoigne received a commission as Providore for Yorkshire in the army of King Charles I. Crabtree lived in Salford, just outside Manchester both of which were on the parliamentary side and all correspondence between the two ceased.

Gascoigne died at the Battle of Marston Moor, Yorkshire, on 2nd July 1644 as did Charles Towneley, the father of his friend Richard Towneley.[1]

After Gascoigne's death some of his papers and fragments of correspondence between Crabtree and Gascoigne came into the possession of Christopher and Richard Towneley. They brought them to the attention of John Flamsteed, the first Astronomer Royal, who came to see Horrocks, Crabtree, and Gascoigne as the founding fathers of British research astronomy and the intellectual heirs of Galileo and Kepler. He began his massive three-folio volume *Historia Coelestis Britannica* (1725) by printing five pages of their surviving letters and observations, made between 1638 and 1643.[4]

Many of Gascoigne's papers and correspondence were lost during the English Civil War and later in the Great Fire of London, but most of what is known to remain is kept in the Bodleian Library at the University of Oxford.[5]

References

[1] http://www.journals.cambridge.org/production/action/cjoGetFulltext?fulltextid=303461 Retrieved 2007-11-24
[2] http://solarsystem.nasa.gov/deepimpact/collaborative_ed_module/4clarify/STWhatGoesAround.pdf Retrieved 2007-11-24
[3] http://www.rod.beavon.clara.net/leonardo.htm Retrieved 2007-11-24
[4] http://www.journals.cambridge.org/production/action/cjoGetFulltext?fulltextid=303461 P. 12 Retrieved 2007-11-25
[5] Catalogue of the scientific and antiquarian papers of Christopher and Richard Towneley, 17th-18th cent. Bodleian Library, University of Oxford (http://www.bodley.ox.ac.uk/dept/scwmss/wmss/online/1500-1900/towneley/towneley000.html)

Further reading

- Chapman, A. *Transits of Venus, History Results and Legacy* (http://www.journals.cambridge.org/production/action/cjoGetFulltext?fulltextid=303461) (pdf file)
- Goward, K. *Jeremiah Horrocks and the Transit of Venus* (http://www.ast.cam.ac.uk/~ipswich/Observations/ToV/Horrocks.htm)
- Sellers, David (2006). *A Letter from William Gascoigne to Sir Kenelm Digby (Journal for the History of Astronomy, v.37, pt.4, Nov 2006, pp.405-416).* Cambridge, England: Science History Publications Ltd. ISSN 0021-8286.

Henry_Gellibrand

Henry Gellibrand (1597–1637) was an English mathematician. He is known for his work on the Earth's magnetic field. He discovered that magnetic declination – the angle of dip of a compass needle – is not constant but changes over time.[1] [2] He announced this in 1635, relying on previous observations by others, which had not yet been correctly interpreted.[3]

He also devised a method for measuring longitude, based on eclipses.[4] The mathematical tables of Henry Briggs, consisting of logarithms of trigonometric functions, were published by Gellibrand in 1633 as *Trigonometria Britannica.*

He was Professor at Gresham College, succeeding Edward Gunter in 1626. He was buried in St Peter Le Poer.[5]

See also

- History of geomagnetism
- List of geophysicists

References

[1] van Helden, Albert; Burr, Elizabeth (1995). "Gellibrand, Henry" (http://galileo.rice.edu/Catalog/NewFiles/gelibrnd.html). *The Galileo Project.* Rice University. . Retrieved 24 October 2011.
[2] http://www.phy6.org/earthmag/mill_3.htm
[3] Lloyd Arnold Brown, The Story of Maps (1979), p. 133.
[4] Michael Hoskin, *The Cambridge Concise History of Astronomy* (2003), p. 150.
[5] http://www.dcs.warwick.ac.uk/bshm/zingaz/LondonPeopleD.html#gun

- Hunter, John Aubrey (2000). "Henry Gellibrand". *Brief lives.* Selected and edited with introduction, glossary and notes by John Buchanan-Brown. Foreword by Michael. Penguin Books. ISBN 978-0140435894.
- "Henry Gellibrand" (http://www-groups.dcs.st-and.ac.uk/~history/Biographies/Gellibrand.html). *The MacTutor History of Mathematics archive.* School of Mathematics and Statistics, University of St. Andrews, Scotland. Retrieved 24 October 2011.

Gerland_(mathematician)

Gerland was an 11th-century mathematician who flourished in England following the Norman Conquest. Little is known of his personal life. His treatise on the Computus may be found in the British Museum and his work on the abacus is preserved in the Bibliothèque nationale de France in Paris.

References

- Wright, Thomas (1846). *Biographia Britannica Literaria: Or, Biography of Literary Characters of Great Britain and Ireland.* J.W. Parker. pp. 16–18.

Peter_Giblin

Peter Giblin	
Nationality	English
Occupation	University Professor
Known for	Work on Singularity Theory

Peter John Giblin is an English mathematician whose primary research involves singularity theory and its application to geometry, computer vision, and computer graphics.[1] Giblin is an emeritus professor of mathematics at the University of Liverpool[2] where he has served on staff for more than 40 years. His positions at Liverpool have included Head of Department (of mathematical sciences), and Head of Division (of pure mathematics).

He is the author or co–author of eight published books, some of which have been translated into Russian. The foreword for the Russian translation of his book *Curves and Singularities* was written by V. I. Arnold. Giblin has also authored or co–authored over a hundred peer reviewed published articles.[3] The first of these was published in 1968.

Publications

- Giblin, P. J. (1977), *Graphs, Surfaces and Homology*, Chapman and Hall, ISBN 0470989947
- Bruce, J. W.; Giblin, P. J. (1984), *Curves and Singularities*, Cambridge University Press, ISBN 0521429994
- Bruce, J. W.; Giblin, P. J.; Rippon, P. J. (1990), *Microcomputers and Mathematics*, Cambridge University Press, ISBN 0521375150
- Giblin, P. J.; Porteus, I. R. (1990), *Challenging Mathematics*, Oxford University Press, ISBN 0199143617
- Giblin, P. J. (1993), *Primes and Programming: an Introduction to Number Theory with Computing*, Cambridge University Press, ISBN 0521401828
- Chen, K.; Giblin, P. J.; Alan Irving, A. (1999), *Mathematical Explorations in Matlab*, Cambridge University Press, ISBN 9780521639200
- Hallinan, P. W.; Gordon, G.; Yuille, A. L.; Giblin, P. J.; Mumford, D. (1999), *Two and Three Dimensional Patterns of the Face*, A. K. Peters, ISBN 978-1-56881-087-4
- Cipolla, R.; Giblin, P. J. (1999), *Visual Motion of Curves and Surfaces*, Cambridge University Press, ISBN 052163251X

Notes and references

[1] Giblin's University of Liverpool homepage (http://www.liv.ac.uk/~pjgiblin/)
[2] University of Liverpool Committee Secretariat Calendar 2009/10 (http://www.liv.ac.uk/commsec/calendar/Emeritus_Professors_2009-102.pdf)
[3] Giblin, Peter. "Bibliography" (http://www.liv.ac.uk/~pjgiblin/pubs/index.html). .

Richard_D._Gill

Richard David Gill	
 Richard D. Gill	
Born	September 11, 1951United Kingdom
Citizenship	British
Alma mater	University of Cambridge Free University of Amsterdam - PhD

Richard David Gill (born 11 September 1951) is a mathematician born in the United Kingdom who has lived in the Netherlands since 1974. As a probability theorist and statistician, Gill is known for his research on counting processes and survival analysis, some of which has appeared in an advanced textbook. He is the chair of mathematical statistics at Leiden University. Gill is also known for his pro bono consulting and advocacy on behalf of victims of incompetent statistical testimony, including a Dutch nurse who was wrongfully convicted and jailed for six years.

Biography

He studied mathematics at the University of Cambridge (1970–1973), and subsequently followed the Diploma of Statistics course there (1973–1974).

Marrying a Dutch woman, he moved to the Netherlands where he worked from 1974 to 1988 at the Mathematical Centre (later renamed *Centrum Wiskunde & Informatica*, or CWI) of Amsterdam. In 1979, Gill obtained his PhD with the thesis *Censoring and Stochastic Integrals*, which was supervised by Jacobus Oosterhoff of the Vrije Universiteit, which awarded the doctorate.[1] Gill spent Autumn 1980 at the Statistical Research Unit at the University of Copenhagen. Gill continued to collaborate with Danish (and Norwegian) statisticians for ten years, helping to write the book *Statistical models based on counting processes*, which is often referred to as "ABGK" (for the authors Andersen, Borgan, Gill, and Keiding).[2] In 1983 he became the head of the Department of Mathematical Statistics at CWI.

In 1988 he moved to the Department of Mathematics of Utrecht University. Gill became the chair in mathematical stochastics—this chair represented the three mathematical sciences of mathematical statistics, probability theory, and operations research. His Ph.D. students include Sara van de Geer and Mark van der Laan.[1]

In 2006, he moved to the Department of Mathematics at Leiden University, where he became the chair of mathematical statistics. Since then, he has conducted statistical research in the theory of quantum information and in biostatistics. He has also worked on survival analysis, semiparametric models, causality, missing data, machine learning, and statistics in image analysis. Gill also publishes on the foundations of several mathematical sciences: the foundations of statistics, of probability, of mathematics, and of quantum physics.

Statistical advocacy against wrongful convictions

In recent years he has actively lobbied for retrials for Lucia de Berk and Kevin Sweeney. The nurse Lucia de Berk was sentenced to life imprisonment following testimony from a legal psychologist that de Berk with the fantastical likelihood committed a string of murders.

> "The court was told by Dr Henk Elffers of the Netherlands Institute for the Study of Crime and Law Enforcement that more children had died on her shifts than appeared possible by chance. He put the odds of her presence being a mere coincidence at one in 342 million, a figure that seemed to have blinded the court to any alternative explanation of the deaths."[3]

This statistical testimony was shown to be fallacious by professional statisticians, notably Gill. Continued scrutiny showed that the data had also been collected to support the prosecutor's conviction of Berk, which further invalidated the pseudo-statistical testimony.[4] [5] [6]

> "The conduct of the case, in Professor Gill's account, was extraordinary. Convinced she was guilty, the police and the managers of the Juliana Children's Hospital assembled a dossier in which it seemed every death became unnatural when it had occurred during, or after, a shift in which she had worked. For one of the alleged murders, it was established on appeal she had not even been in the hospital for three days around the time it occurred. Using more appropriate statistical methods reduced the odds from one in 342 million to one in 48. A further analysis by Professor Gill further reduces the odds to one in nine."[3]

Professor Gill helped in the campaign to have a new trial. Consequently, a retrial was ordered, and de Berk was found not guilty, and received a public apology from the Dutch government, along with financial compensation (amount unknown) for her six years of incarceration.[7] [8]

Honours

Richard Gill is a member of the Royal Netherlands Academy of Arts and Sciences.[9] He is the current Board Chairman for The Netherlands Society for Statistics and Operations Research, which publishes the journal *Statistica Neerlandica*.[10] Gill has been selected as the 2010-2011 Distinguished Lorentz Fellow [11] by the Netherlands Institute for Advanced Study in Humanities and Social Sciences.[12]

References

[1] Richard David Gill (http://genealogy.math.ndsu.nodak.edu/id.php?id=35026) at the Mathematics Genealogy Project.

[2] Andersen, Per Kragh; Borgan, Ørnulf; Gill, Richard D.; Keiding, Niels (1993). *Statistical models based on counting processes*. Springer series in statistics. New York: Springer-Verlag. pp. xii+767. ISBN 0-387-97872-0. MR1198884.

[3] Hawkes, Nigel (*Saturday, 10 April 2010*). "Did statistics damn Lucia de Berk?: Behind the numbers" (http://www.independent.co.uk/opinion/commentators/nigel-hawkes-did-statistics-damn-lucia-de-berk-1940735.html). *The Independent (London)*. .

[4] Mark Buchanan (18 January 2007). "Statistics: conviction by numbers" (http://www.nature.com/nature/journal/v445/n7125/pdf/445254a.pdf). *Nature* **445** (7125): 254–255. Bibcode 2007Natur.445..254B. doi:10.1038/445254a. PMID 17230166. .

[5] Persbericht CWI (http://www.cwi.nl/pr/press-releases/2007/pb-petitieGrunwald-011107.html) Petitie 2 november 2007

[6] "Expert on the most important proof in the Lucia de B. case: 'This baby has not been poisoned'". (http://www.novatv.nl/index.cfm?ln=nl&fuseaction=artikelen.details&achtergrond_id=9907) *NOVA*. September 29, 2007.

[7] "Nurse Lucia de Berk finally found not guilty of murdering seven patients" (http://www.dutchnews.nl/news/archives/2010/04/nurse_lucia_de_berk_not_guilty.php). 14 April 2010. .

[8] "Apology for nurse jailed for murdering seven patients" (http://www.independent.co.uk/news/world/europe/dutch-court-acquits-lucia-de-berk-of-killing-patients-1944577.html), AP, *The Independent* 14 April 2010.

[9] Gill, Prof. dr. R.D., Member info, (http://www.knaw.nl/cfdata/members/list.cfm?persoon=4165) Royal Netherlands Academy of Arts and Sciences. Accessed january 23, 2010

[10] Board (http://www.vvs-or.nl/english/indexenglish.htm), Netherlands Society for Statistics and Operations Research. Accessed January 23, 2010

[11] http://www.nias.knaw.nl/en/fellowships/distinguished_lorentz_fellowship/

[12] Richard Gill Distinguished Lorentz Fellow 2010-2011 (http://www.nieuws.leidenuniv.nl/laureaten/richard-gill-distinguished-lorentz-fellow-2010-2011.html), News release, Leiden University. Accessed January 23, 2010.

External links

- Richard Gill's homepage at Leiden University (http://www.math.leidenuniv.nl/~gill).
- Richard D. Gill (http://genealogy.math.ndsu.nodak.edu/id.php?id=35026) at the Mathematics Genealogy Project.
- Mathematical Reviews. "Richard D. Gill" (http://www.ams.org/mathscinet/search/author.html?mrauthid=73585). Retrieved 2011-02-22.

James_Whitbread_Lee_Glaisher

James Whitbread Lee Glaisher FRS FRAS (5 November 1848 – 7 December 1928) son of James Glaisher, the meteorologist, was a prolific English mathematician and astronomer.

James Whitbread Lee Glaisher.

He was educated at St Paul's School and Trinity College, Cambridge, where he was second wrangler in 1871.[1] Influential in his time on teaching at the University of Cambridge, he is now remembered mostly for work in number theory that anticipated later interest in the detailed properties of modular forms. He published widely over other fields of mathematics.

He was the editor-in-chief of *Messenger of Mathematics*.

See also

- Glaisher's theorem
- Glaisher–Kinkelin constant

References

[1] James Whitbread Lee Glaisher (http://venn.lib.cam.ac.uk/cgi-bin/search.pl?sur=&suro=c&fir=&firo=c&cit=&cito=c&c=all&tex=GLSR866JW&sye=&eye=&col=all&maxcount=50) in Venn, J. & J. A., *Alumni Cantabrigienses*, Cambridge University Press, 10 vols, 1922–1958.

External links

- Works written by or about James Whitbread Lee Glaisher at Wikisource
- O'Connor, John J.; Robertson, Edmund F., "James Whitbread Lee Glaisher" (http://www-history.mcs.st-andrews.ac.uk/Biographies/Glaisher.html), *MacTutor History of Mathematics archive*, University of St Andrews.
- Univ. of St. Andrews biography (http://www-groups.dcs.st-and.ac.uk/~history/Biographies/Glaisher.html)

Alfred_Goldie

Alfred William Goldie (December 10, 1920, Coseley, Staffordshire – October 8, 2005) was an English mathematician.

Goldie was Assistant Lecturer, Nottingham University 1946-48; Lecturer, then Senior Lecturer, Newcastle University 1948-63; and Professor of Pure Mathematics, Leeds University 1963-86;

He won the 1970 Senior Berwick Prize from the London Mathematical Society, where he also became Vice-President from 1978-80.

He worked in ring theory where he introduced the notion of the uniform dimension of a module, and the reduced rank of a module. He is well known for Goldie's theorem, which characterizes right Goldie rings.

References

- The Independent Obituaries: Alfred W. Goldie, Algebraist 'Lord of the Rings' [1]
- Leeds Campusweb | Obituaries news item: Emeritus Professor Alfred Goldie, MA [2]
- Mathematics People, Notices AMS [3]
- S C Coutinho and J C McConnell, The Quest for Quotient Rings (Of Noncommutative Noetherian Rings) [4], Amer. Math. Monthly 110 (4) (2003), 298-313.

References

[1] http://www.independent.co.uk/news/obituaries/alfred-w-goldie-521319.html

[2] http://campus.leeds.ac.uk/newsincludes/newsitem3269.htm

[3] http://www.ams.org/notices/200601/people.pdf

[4] http://www.dcc.ufrj.br/~collier/artigos/Quest.pdf

Benjamin_Gompertz

Benjamin Gompertz	

Born	5 March 1779London, England
Died	14 July 1865 (aged 86)
Nationality	Great Britain
Fields	Mathematics

Benjamin Gompertz (5 March 1779 – 14 July 1865) was a British self educated mathematician and actuary, who became a Fellow of the Royal Society. Gompertz is now best known for his Gompertz law of mortality, a demographic model published in 1825.

Life

Of the German Jewish family of Gompertz of Emmerich, he was born in London, where his father and grandfather had been successful diamond merchants. Debarred, as a Jew, from a university education, he studied on his own from an early age, in the writings of Isaac Newton, Colin Maclaurin, and William Emerson. From 1798 he was a prominent contributor to the *Gentleman's Mathematical Companion*, and for a period won the annual prizes in the magazine for the solutions of problems.[1]

In line with his father's wishes, he entered the London Stock Exchange. He became a member of the Mathematical Society of Spitalfields, and served as its president when it was merged with the Astronomical Society of London. In 1819 he was elected a F.R.S., and in 1832 became a member of the council. The Astronomical Society was founded in 1820, and he was elected a member of the council in 1821.[1]

On the death of his only son he retired from the Stock Exchange, and absorbed himself in mathematics. When the Guardian Insurance Office was established in 1821, he was a candidate for the actuaryship, but the directors objected to him on the grounds that he was a Jew. His brother-in-law, His brother-in-law Sir Moses Montefiore with his relative Nathan Mayer Rothschild then founded the Alliance Assurance Company (1824), and Gompertz was appointed actuary under the deed of settlement. His management of the Alliance Company was successful, he was consulted by government, and made computations for the Army medical board.[1]

In 1848 he retired from active work and returned to his scientific labours. He was a member of many learned societies, and was also one of the promoters of the Society for the Diffusion of Useful Knowledge. Of the leading Jewish charities he was a prominent member, and he worked out a plan of poor relief (*Jewish Chronicle*, 6 October 1845), which was afterwards adopted by the Jewish board of guardians.[1]

Gompertz died from a paralytic seizure on 14 July 1865.[1]

Works

Mathematics

From 1806 he was a frequent contributor to the *Philosophical Transactions* of the Royal Society; but his early tracts on complex numbers and porisms (1817–18) were self-published. Gompertz was an old-fashioned Newtonian who retained and defended the notation of fluxions.[1]

Astronomy

For ten years he actively participated in the work of the Astronomical Society, contributing papers on the theory of astronomical instruments, the aberration of light, the differential sextant, and the convertible pendulum. With Francis Baily he began in 1822 the construction of tables for the mean places of the fixed stars; the work was left uncompleted, because of the publication of the *Fundamenta Astronomiæ* of Friedrich Bessel. Their efforts, however, led to the complete catalogue of stars of the Royal Astronomical Society.[1]

The Gompertz model

He worked out a new series of tables of mortality for the Royal Society, and these suggested to him in 1825 his law of human mortality, which he first expounded in a letter to Francis Baily. The law rests on an *a priori* assumption that a person's resistance to death decreases as his years increase. The model can be written in this way:

where $N(t)$ represents the number of individuals at time t, and c and a are constants.

This model is a refinement of a demographic model of Robert Malthus. It was used by insurance companies to calculate the cost of life insurance. The equation, known as a Gompertz curve, is now used in many areas to model a time series where growth is slowest at the start and end of a period. The model has been extended to the Gompertz–Makeham law of mortality.

Family

He married Abigail Montefiore in 1810.[1]

Gompertz's descendant Simon Gompertz is an economics presenter for the BBC.

See also

- Population dynamics

References

- "On the Nature of the Function Expressive of the Law of Human Mortality, and on a New Mode of Determining the Value of Life Contingencies" [2]. *Philosophical Transactions of the Royal Society of London* **115**: 513–585. 1825. doi:10.1098/rstl.1825.0026.

Notes

[1] "Gompertz, Benjamin". *Dictionary of National Biography*. London: Smith, Elder & Co. 1885–1900.
[2] http://visualiseur.bnf.fr/Visualiseur?Destination=Gallica&O=NUMM-55920

External links

- O'Connor, John J.; Robertson, Edmund F., "Benjamin Gompertz" (http://www-history.mcs.st-andrews.ac.uk/Biographies/Gompertz.html), *MacTutor History of Mathematics archive*, University of St Andrews.
- P.F. Hooker - Benjamin Gompertz, 5 March 1779 - 14 July 1865, 1965 (http://www.actuaries.org.uk/files/pdf/library/JIA-091/0203-0212.pdf)

Attribution

I._J._Good

Irving John ("I.J.") Good	
Born	9 December 1916London, England, UK
Died	April 5, 2009 (aged 92)Radford, Virginia, USA
Fields	Statistician, cryptologist
Institutions	Trinity College, Oxford; Virginia Tech
Alma mater	Jesus College, Cambridge
Doctoral advisor	G. H. Hardy

Irving John ("I.J."; "Jack") Good (9 December 1916 – 5 April 2009)[1] [2] was a British mathematician who worked as a cryptologist at Bletchley Park with Alan Turing. After World War II, Good continued to work with Turing on the design of computers and Bayesian statistics at the University of Manchester. Good moved to the United States where he was professor at Virginia Tech.

He was born **Isadore Jacob Gudak** to a Polish-Jewish family in London. He later anglicized his name to Irving John Good and signed his publications "**I. J. Good**."

An originator of the concept now known as "technological singularity," Good served as consultant on supercomputers to Stanley Kubrick, director of the 1968 film *2001: A Space Odyssey*.

Life

Good was born Isadore Jacob Gudak to Polish-Jewish parents in London. His father was a watchmaker, who later managed and owned a successful fashionable jewelry shop, and was also a notable Yiddish writer writing under the pen-name of Moshe Oved. Good was educated at the Haberdashers' Aske's boys' school in Hampstead, north London, where, Dan van der Vat writes, Good effortlessly outpaced the mathematics curriculum.[3]

Good studied mathematics at Jesus College, Cambridge, graduating in 1938 and winning the Smith's Prize in 1940. He did research under G.H. Hardy and Besicovitch before moving to Bletchley Park in 1941 on completing his doctorate.

Bletchley Park

On 27 May 1941, having just obtained his doctorate at Cambridge, Good walked into Hut 8, Bletchley's facility for breaking German naval ciphers, for his first shift. This was the day that Britain's Royal Navy destroyed the German battleship *Bismarck* after it had sunk the Royal Navy's *HMS Hood*. Bletchley had contributed to *Bismarck's* destruction by discovering, through wireless-traffic analysis, that the German flagship was sailing for Brest, France, rather than Wilhelmshaven, from which she had set out.[3] Hut 8 had not, however, been able to decrypt on a current basis the 22 German Naval Enigma messages that had been sent to *Bismarck*. The German Navy's Enigma ciphers were considerably more secure than those of the German Army or Air Force, which had been well penetrated by 1940. Naval messages were taking three to seven days to decrypt, which usually made them operationally useless for the British. This was about to change, however, with Good's help.[3]

> Alan Turing... had caught Good sleeping on the floor while on duty during his first night shift. At first, Turing thought Good was ill, but he was cross when Good explained that he was just taking a short nap because he was tired. For days afterwards, Turing would not deign to speak to Good, and he left the room if Good walked in. The new recruit only won Turing's respect after he solved the bigram tables problem. During a subsequent night shift, when there was no more work to be done, it dawned on Good that there might be another chink in

> the German indicating system. The German telegraphists had to add dummy letters to the trigrams which they selected out of the *kenngruppenbuch*... Good wondered if their choice of dummy letters was random, or whether there was a bias towards particular letters. After inspecting some messages which had been broken, he discovered that there was a tendency to use some letters more than others. That being the case, all the codebreakers had to do, was to work back from the indicators given at the beginning of each message, and apply each bigram table in turn in the same way as Joan Clarke had done before. The bigram table which produced one of the popular dummy letters was probably the correct one. When Good mentioned his discovery to Alan Turing, Turing was very embarrassed, and said, 'I could have sworn that I tried that.' It quickly became an important part of the Banburismus procedure.
>
> Jack Good's refusal to go on working when tired was vindicated by a subsequent incident. During another long night shift, he had been baffled by his failure to break a doubly enciphered *Offizier* message. This was one of the messages which was supposed to be enciphered initially with the Enigma set up in accordance with the *Offizier* settings, and subsequently with the general Enigma settings in place. However, while he was sleeping before returning for another shift, he dreamed that the order had been reversed; the general settings had been applied before the *Offizier* settings. Next day he found that the message had yet to be read, so he applied the theory which had come to him during the night. It worked; he had broken the code in his sleep.[4]

Good served with Turing for nearly two years.[3]

Subsequently he worked with Donald Michie in Max Newman's group on the Fish ciphers, leading to the development of the Colossus computer.

Postwar work

In 1947 Newman invited Good to join him and Turing at Manchester University. There for three years Good lectured in mathematics and researched computers, including the Manchester Mark 1.[3]

In 1948 Good was recruited by the Government Communications Headquarters (GCHQ), successor to Bletchley Park. He remained there until 1959, while also taking up a brief associate professorship at Princeton University and a short consultancy with IBM.[3]

From 1959 until he moved to the U.S. in 1967, Good held government-funded positions and from 1964 a senior research fellowship at Trinity College, Oxford, and the Atlas Computer Laboratory, where he continued his interests in computing, statistics and chess.[2] He later left Oxford, declaring it "a little stiff".

United States

In 1967 Good moved to the United States, where he was appointed a research professor of statistics at Virginia Polytechnic Institute and State University. In 1969 he was appointed a University Distinguished Professor at Virginia Tech, and in 1994 Emeritus University Distinguished Professor.[5]

Here's something he had later said about his arrival in Virginia (from England) in 1967 to start teaching at VPI, where he taught from 1967 to 1994:

"I arrived in Blacksburg in the seventh hour of the seventh day of the seventh month of the year seven in the seventh decade, and I was put in Apartment 7 of Block 7...all by chance."

Research and publications

Good's published work ran to over three million words.[3] He was known for his work on Bayesian statistics. He published a number of books on probability theory. In 1958 he published an early version of what later became known as the Fast Fourier Transform[6] but in a journal so obscure that it never became widely known. He played chess to county standard and helped popularize Go, an Asian boardgame, through a 1965 article in *New Scientist* (he had learned the rules from Alan Turing).[7] In 1965 he originated the concept now known as "technological singularity," which anticipates the eventual advent of superhuman intelligence:

> Let an ultraintelligent machine be defined as a machine that can far surpass all the intellectual activities of any man however clever. Since the design of machines is one of these intellectual activities, an ultraintelligent machine could design even better machines; there would then unquestionably be an 'intelligence explosion,' and the intelligence of man would be left far behind. Thus the first ultraintelligent machine is the last invention that man need ever make. [8]

Good's authorship of treatises such as "Speculations Concerning the First Ultraintelligent Machine" and "Logic of Man and Machine" (both 1965) made him the obvious person for Stanley Kubrick to consult when filming *2001: A Space Odyssey* (1968), one of whose principal characters was the paranoid HAL 9000 supercomputer.[3] In 1995 Good was elected a member of the Academy of Motion Picture Arts and Sciences.[2]

Personality

The slender, bushy-moustached Good was blessed with a sense of humor. He published a paper under the names IJ Good and "K Caj Doog"—the latter, his own nickname spelled backwards. In a 1988 paper,[9] he introduced its subject by saying, "Many people have contributed to this topic but I shall mainly review the writings of I. J. Good because I have read them all carefully." In Virginia he chose, as his vanity license plate, "007IJG," in subtle reference to his World War II intelligence work.[3]

Death

Good died on 5 April 2009 of natural causes in Radford, Virginia, aged 92.[10]

Books

- *Probability and the Weighing of Evidence* (1950), Griffin, London.
- *Information, Weight of Evidence: The Singularity Between Probability Measures and Signal Detection* (1974) with D.B. Osteyee, Springer, ISBN 978-3540067269.
- *Good Thinking: The Foundations of Probability and Its Applications* (1983) University of Minnesota Press. Republished by Dover in 2009.

See also

- Good-Turing frequency estimation
- Cryptanalysis of the Enigma
- MacMahon Master theorem

Notes

[1] "Passings" (http://www.latimes.com/news/obituaries/la-me-passings13-2009apr13,0,3844936.story). *Los Angeles Times*. April 13, 2009. . Retrieved April 13, 2009.

[2] The Times of 16-apr-09, http://www.timesonline.co.uk/tol/comment/obituaries/article6100314.ece

[3] Dan van der Vat, "Jack Good" (obituary), *The Guardian*, 29 April 2009, p. 32.

[4] Hugh Sebag-Montefiore, *Enigma: The Battle for the Code*, p. 189.

[5] "Good, Irving John" (http://www.stat.vt.edu/facstaff/IJGoodVitae.pdf). *CV*. Virginia Polytechnic Institute and State University. July 24 04. . Retrieved 4 October 2010.

[6] "The Interaction Algorithm and Practical Fourier Analysis," J. Royal Statistical Society (1958), Ser.B., Vol.20,pp.361-372.

[7] "The mystery of Go," *The New Scientist*, January 1965, pp. 172-74.

[8] I.J. Good, "Speculations Concerning the First Ultraintelligent Machine" (http://commonsenseatheism.com/wp-content/uploads/2011/02/Good-Speculations-Concerning-the-First-Ultraintelligent-Machine.pdf) (HTML (http://www.acceleratingfuture.com/pages/ultraintelligentmachine.html)), *Advances in Computers*, vol. 6, 1965.

[9] I.J. Good, "The Interface Between Statistics and Philosophy of Science," (http://links.jstor.org/sici?sici=0883-4237(198811)3:4<386:TIBSAP>2.0.CO;2-I) *Statistical Science*, vol. 3, no. 4, 1988, pp. 386–97.
[10] Virginia Tech news release of Good's death. (http://www.vtnews.vt.edu/story.php?relyear=2009&itemno=276)

References

- Dan van der Vat, "Jack Good" (obituary), *The Guardian*, 29 April 2009, p. 32.
- Hugh Sebag-Montefiore, *Enigma: The Battle for the Code*, London, Weidenfeld & Nicolson, 2000, ISBN 978-0297842514.

External links

- I. J. Good (http://genealogy.math.ndsu.nodak.edu/id.php?id=72215) at the Mathematics Genealogy Project.
- Good's web page (http://www.stat.vt.edu/facstaff/ijgood.html) at Virginia Tech
- Bibliography ("Shorter Publications List", running to 2300 items) (PDF) (http://www.web-e.stat.vt.edu/holtzman/IJGood/IJGoodsShorterPubList,Aug19,2003.pdf)
- Biography (http://ei.cs.vt.edu/~history/Good.html) focusing on Good's role in the history of computing
- Project Euclid (http://projecteuclid.org/Dienst/UI/1.0/Summarize/euclid.ss/1032209661) An interview with Good can be downloaded from here
- VT Image Base (http://imagebase.lib.vt.edu/browse.php?folio_ID=/va/fac/good) Photographs
- Obituary (http://www.vtnews.vt.edu/story.php?relyear=2009&itemno=276), Virginia Tech, 6 April 2009
- Obituary (http://www.telegraph.co.uk/news/obituaries/military-obituaries/special-forces-obituaries/5132599/Professor-Jack-Good.html), *Daily Telegraph*, 10 April 2009
- Obituary (http://www.timesonline.co.uk/tol/comment/obituaries/article6100314.ece), *The Times*, 16 April 2009
- Obituary (http://www.independent.co.uk/news/obituaries/jack-good-cryptographer-whose-work-with-alan-turing-at-bletchley-park-was-crucial-to-the-war-effort-1684506.html), *The Independent*, 14 May 2009
- Eulogy (http://www.math.rutgers.edu/~zeilberg/mamarim/mamarimhtml/jack.html) Mathematical eulogy (with Maple code) by Doron Zeilberger, 2 December 2009

Reuben_Goodstein

Reuben Louis Goodstein (15 December 1912 in London – 8 March 1985 in Leicester) was an English mathematician with a strong interest in the philosophy and teaching of mathematics.

As a boy, he attended St Paul's School in London. He received his Master's degree from the University of Cambridge. After this, he worked at the University of Reading but ultimately spent most of his academic career in the University of Leicester. He earned his PhD from the University of London in 1946 while still working in Reading. Goodstein also studied under Wittgenstein and John Littlewood.

He published many works on finitism and the reconstruction of analysis from a finitistic viewpoint, for example "Constructive Formalism. Essays on the foundations of mathematics." Goodstein's theorem was among the earliest examples of theorems found to be unprovable in Peano arithmetic but provable in stronger logical systems (such as second order arithmetic). He also introduced a variant of the Ackermann function that is now known as the hyperoperation sequence, together with the naming convention now used for these operations (*tetration*, *pentation*, etc.).

See also

- Goodstein's theorem
- Primitive recursive arithmetic

External links

- O'Connor, John J.; Robertson, Edmund F., "Reuben Goodstein" [1], *MacTutor History of Mathematics archive*, University of St Andrews.

References

[1] http://www-history.mcs.st-andrews.ac.uk/Biographies/Goodstein.html

John_Gough_(natural_philosopher)

John Gough	
Born	January 15, 1757Kendal, Westmorland, England
Died	July 28, 1825 (aged 68)Kendal, Westmorland, England
Residence	England
Nationality	English
Fields	Polymath, natural philosopher
Notable students	John Dalton William Whewell Richard Dawes Thomas Gaskin

John Gough (17 January 1757 – 28 July 1825) was a blind English natural and experimental philosopher who is known for his own investigations as well as the influence he had on both John Dalton and William Whewell.

Life

John Gough was born in Kendal, Westmorland, on 17 January 1757, the eldest child of Nathan Gough (d. 1800) and his wife, Susannah (1731 – 1798). Gough's father was a wool dyer and shearman dyer,[1] while his mother was the eldest daughter of John Wilson, a prosperous farmer with an estate on the west bank of Windermere. Nathan and Susannah Gough had three sons and four daughters, one of whom died in infancy. The family belonged to the Society of Friends, whose communities flourished in Cumberland and Westmorland during this period. Before he was three years old, Gough was attacked by smallpox and lost his sight. In his childhood he expended much effort in developing his sense of touch and hearing, and appears to have been especially eager to learn to recognize animals by touch.

In 1778 at the age of eighteen, Gough became a resident pupil of John Slee, a mathematical master at Mungrisedale, Cumberland. Gough stayed at Mungrisedale for eighteen months, following the traditional curriculum up to the elementary principles of calculus. Returning home he took up calculus, with his second sister, Dorothy Gough (b. 1768), acting as his reader. From around 1782 to 1790 he enjoyed the acquaintance of John Dalton, a cousin of George Bewley and also a lakeland Quaker, who had come to Kendal to take up a position in Bewley's school. Dalton assisted Gough by reading, writing, and making calculations and diagrams on his behalf. In return Dalton, who later became one of the most eminent figures in nineteenth-century science, was tutored by Gough in Latin and Greek. Dalton later referred to Gough as a "prodigy in scientific attainments."[2]

In 1800 at the parish church of Kendal, Gough married Mary (d. 1858), daughter of Thomas Harrison of Crosthwaite, Cumberland. On their marriage they moved to Middleshaw in the hamlet of Old Hutton. John and Mary Gough had nine children, one of whom, Thomas Gough (1804 – 1880) became a surgeon in Kendal.

In 1812 Gough had a house, which he named Fowl Ing, built for himself and his family on the south-west slope of Benson Knot, a hill 2 miles north-east of Kendal. At about the same time he began to act as a private tutor of mathematics to a select group of pupils from northern England, whom he prepared for university. The subsequent fame of his students superseded his own celebrity. A number of them went on to achieve high distinction in the mathematical tripos, and subsequently in the hierarchies of university and church. One of Gough's first students was William Whewell, who was with him in 1812 and later described Gough as "a very extraordinary person."[3]

From 1823, John Gough suffered repeated attacks of epilepsy. He died on 28 July 1825, leaving his wife and seven of their children. He was buried in the parish churchyard of Kendal.

Career

Gough had wide-ranging scientific interests. He published papers in natural history, mechanics, mathematics, chemistry, and experimental physics. One of his most interesting pieces of work was an investigation into the properties of natural rubber or Caoutchouc. He was the first to describe the heat released when a rubber band is quickly stretched, the heat being detected by the lips to which the band is pressed. When stretched rubber is heated, it contracts, a reversal of the normal behaviour of materials when heated. Gough published these results, and others, in a letter to the Manchester Literary and Philosophical Society in 1804. King has written[4] that through Dalton, Gough exerted an indirect influence years later on James Joule, who undertook his own investigations into rubber, elasticity, and energy changes, and specifically referred to Gough's earlier studies.[5]

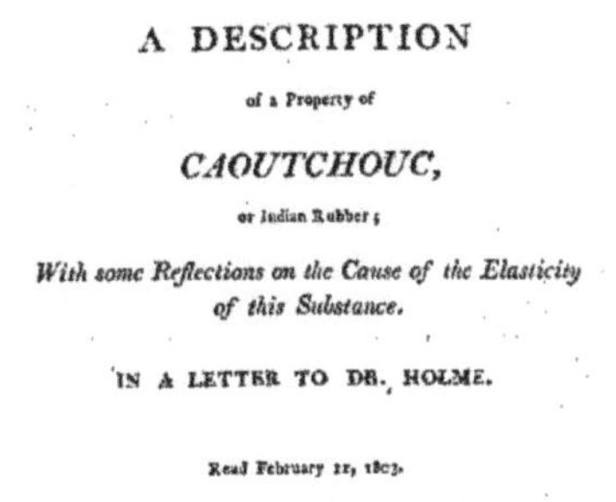
A DESCRIPTION
of a Property of
CAOUTCHOUC,
or Indian Rubber;
With some Reflections on the Cause of the Elasticity of this Substance.
IN A LETTER TO DR. HOLME.
Read February 11, 1803.

Gough's paper on rubber

Gough's most substantive enquiry was "An investigation of the method whereby men judge by the ear of the position of sonorous bodies relative to their own persons", which appeared in 1802 during an ongoing controversy with another former Quaker, the noted natural philosopher Thomas Young, over the nature of compound sounds.

References

[1] Gough, Thomas (1861), "John Gough" (http://books.google.com/books?id=G78HAAAAQAAJ&printsec=frontcover&dq=annals+of+kendal#PPR1,M2), in Nicholson, Cornelius, *The Annals of Kendal* (2 ed.), London: Whitaker and Company, pp. 355–368, . - See preface for authorship of the article.

[2] Dalton, John (1834), *Meteorological Observations and Essays* (http://books.google.com/?id=Ot8KAAAAIAAJ&printsec=frontcover&dq=Meteorological+observations+and+essays#PPP16,M2) (2 ed.), Manchester: Harrison and Crosfield, pp. xvi,

[3] See page 6 of Douglas, Stair (1881), *The Life and Selections from the Correspondence of William Whewell, D.D. Late Master of Trinity College Cambridge* (http://books.google.com/?id=q0YOAAAAYAAJ&printsec=frontcover&dq=The+life+of+William+Whewell#PPA6,M1), London: C. Kegan Paul and Company,

[4] King (1953), "Note in Memory of Blind John Gough", *American Journal of Physics* **21** (3): 231–232, doi:10.1119/1.1933400.

[5] Joule, James Prescott (1884), *The Scientific Papers of James Joule* (http://books.google.com/?id=UR5WAAAAMAAJ&printsec=frontcover&dq=joule's+scientific+papers), London: Taylor and Francis, pp. 428–440,

Further reading

- Coleridge, Samuel Talor (1812), "The Soul and its Organs of Sense" (http://www.archive.org/details/omnianaorhoraeot02soutrich), in Southey, Robert, *Omniana*, **2**, London: Longman, Hurst, Rees, Orme, and Brown, pp. 17–18
- Curwen, John F. (1900), *Kirkbie-Kendal* (http://books.google.com/?id=ORMPAAAAYAAJ&pg=RA1-PA54&dq=Kirkbie+Kendal#PPR3,M1), Kendal: T. Wilson, pp. 308–309, 349, 407, 424
- Foster, Joseph (1890), *The Pedigree of Wilson of High Wray and Kendal* (http://books.google.com/?id=ilIBAAAAQAAJ&printsec=frontcover&dq=The+Pedigree+of+Wilson+of+High+Wray+and+Kendal#PPP7,M1), pp. 96
- Henry, William Charles (1854), *Memoirs of the Life and Scientific Researches of John Dalton* (http://books.google.com/?id=ItElAAAAMAAJ&printsec=frontcover&dq=Memoirs+of+the+Life+and+Scientific+Researches+of+John+Dalton#PPR3,M1), London: Harrison and Sons, pp. 9–11
- Nicholson, F.; Axon, E. (1915), *The Older Nonconformity in Kendal* (http://www.archive.org/details/oldernonconformi00nichiala), Kendal: Titus Wilson, pp. 374–375, 378, 523–524

External links

- Gough on the thermodynamics of rubber (http://www.chem.yale.edu/~chem125/goodyear/Gough.html)

Timothy_Gowers

Timothy Gowers

Timothy Gowers in 2009

Born	20 November 1963Wiltshire, England
Citizenship	British
Institutions	University of Cambridge University College London
Alma mater	University of Cambridge
Doctoral advisor	Béla Bollobás
Doctoral students	Pablo Candela David Conlon Ben J. Green George Petridis Tom Sanders Mark Walters Julia Wolf András Zsák
Known for	Functional analysis, combinatorics
Notable awards	Prize of the European Mathematical Society (1996) Fields Medal (1998)

William Timothy Gowers FRS (born 20 November 1963, Wiltshire) is a British mathematician. He is a Royal Society Research Professor at the Department of Pure Mathematics and Mathematical Statistics at Cambridge University, where he also holds the Rouse Ball chair, and is a Fellow of Trinity College. In 1998 he received the Fields Medal for his research connecting the fields of functional analysis and combinatorics.

Education

His early education was at King's College School, Cambridge as a chorister in the Choir of King's College, Cambridge, and at Eton College[1] where he was a King's Scholar. He completed his PhD entitled 'Symmetric Structures in Banach Spaces' at the University of Cambridge (Trinity College) in 1990 under the supervision of Béla Bollobás.

Career

From 1991 to 1995 he was a member of the Department of Mathematics at University College London. In 1996 he received the Prize of the European Mathematical Society and in 1998 the Fields Medal for research on functional analysis and combinatorics. He used combinatorial tools in proving several of Stefan Banach's conjectures on Banach spaces and in constructing a Banach space with almost no symmetry, serving as a counterexample to several other conjectures.[2] With Bernard Maurey he resolved the "unconditional basic sequence problem" in 1992, showing that not every infinite-dimensional Banach space has an infinite-dimensional subspace that admits an unconditional Schauder basis. Another work which has proved highly influential is his proof of Szemerédi's theorem by Fourier-analytic methods. He has also made substantial contributions in combinatorics, particularly to the study of regularity for graphs and hypergraphs. In 1999 he was elected Fellow of the Royal Society.

In addition to scholarly papers on mathematics, Gowers is also the author of several works popularizing mathematics, including the 2002 book *Mathematics: A Very Short Introduction*[3] which describes modern research mathematics to the layman. He was consulted about the 2005 film *Proof*, starring Gwyneth Paltrow and Anthony Hopkins. Recently, he has been the editor for *The Princeton Companion to Mathematics*, a voluminous book published in 2008 that introduces and traces the development of various branches and concepts of modern mathematics. For his work on the *Princeton Companion*, Gowers won the 2011 Euler Book Prize of the Mathematical Association of America.[4]

Polymath Project

In a 2009 post on his blog, Gowers asked the provocative question "is massively collaborative mathematics possible?" [5].[6] This post led to his creation of the Polymath Project [7], using the comment functionality of his blog to produce mathematics collaboratively.[8]

The initial proposed problem [9] for this project, now called Polymath1 by the Polymath community, was to find a new combinatorial proof to the density version of the Hales–Jewett theorem. As the project took form, two main threads of discourse emerged. The first thread, which was carried out in the comments of Gowers's blog, would continue with the original goal of finding a combinatorial proof. The second thread, which was carried out in the comments of Terry Tao's blog, focused on calculating bounds on density Hales-Jewett numbers and Moser numbers for low dimensions.

After 7 weeks, Gowers announced on his blog that the problem was "probably solved",[10] though work would continue on both Gowers's thread and Tao's thread well into May 2009, some three months after the initial announcement. In total over 40 people contributed to the Polymath1 project. Both threads of the Polymath1 project have been successful, producing at least two new papers to be published under the pseudonym D.H.J Polymath.[11] [12] [13]

The success of the Polymath1 project has spawned additional Polymath projects. To date there are 5 official Polymath Projects, and 2 Mini-Polymath Projects. More information on the Polymath1 project can be found on the project wiki [14]. Jason Dyer's *A gentle introduction to the Polymath project* [15] provides a good explanation of the work of the project for a non-mathematical audience.

Tricki

Tricki.org [16] is a Wikipedia-style project collecting methods of mathematical problem solving conceived in 2008 and launched by Gowers, Olof Sisask and Alex Frolkin in March 2009.[17] Terence Tao and Ben Green are among those to have contributed articles.[18]

Personal life

He is the son of composer Patrick Gowers, great-grandson of British civil servant Sir Ernest Gowers and great-great-grandson of neurologist Sir William Gowers. He has five children[19] and plays jazz piano.[1] For ten days of every November, Gowers and his five children all have either only odd or only even number ages.

Selected publications

- Gowers, W.T.; Maurey, Bernard (6 May 1992). "The unconditional basic sequence problem.org". arXiv:math/9205204 [math.FA].
- Gowers, W.T. (2001). "A new proof of Szemerédi's theorem". *Geom. Funct. Anal. 11, 465–488.*
- Gowers (2007). "Hypergraph regularity and the multidimensional Szemer\'edi theorem". arXiv:0710.3032v1 [math.CO].
- Gowers, W.T. (2007). "Hypergraph regularity and the multidimensional Szemerédi theorem". *Ann. of Math. 166, 897–946.*
- Gowers, Timothy, ed (2008). *The Princeton Companion to Mathematics*. Princeton University Press. ISBN 978-0691118802.

Notes

[1] Sleeman, Elizabeth (2003). *The International Who's Who 2004*. Routledge. p. 637. ISBN 1857432177.

[2] 1998 Fields Medalist William Timothy Gowers (http://www.ams.org/featurecolumn/archive/gowers.html) from the American Mathematical Society

[3] Gowers, Timothy (August 2002). *Mathematics: A Very Short Introduction*. Oxford Paperbacks. ISBN 978-0192853615.

[4] January 2011 Prizes and Awards (http://www.ams.org/profession/prize-booklet-2011.pdf), American Mathematical Society, retrieved 2011-02-01.

[5] http://gowers.wordpress.com/2009/01/27/is-massively-collaborative-mathematics-possible/

[6] Gowers, T.; Nielsen, M. (2009). "Massively collaborative mathematics". *Nature* **461** (7266): 879–881. Bibcode 2009Natur.461..879G. doi:10.1038/461879a. PMID 19829354.

[7] http://michaelnielsen.org/polymath1/index.php?title=Main_Page

[8] Gowers, Timothy (2009-01-27). *Is massively collaborative mathematics possible?* (http://gowers.wordpress.com/2009/01/27/is-massively-collaborative-mathematics-possible/). Gowers's Weblog. . Retrieved 2009-03-30.

[9] http://gowers.wordpress.com/2009/02/01/a-combinatorial-approach-to-density-hales-jewett/

[10] Nielsen, Michael (2009-03-20). "The Polymath project: scope of participation" (http://michaelnielsen.org/blog/?p=584). . Retrieved 2009-03-30.

[11] Polymath (2010). "Deterministic methods to find primes". arXiv:1009.3956 [math.NT].

[12] Polymath (2010). "Density Hales-Jewett and Moser numbers". arXiv:1002.0374 [math.CO].

[13] Polymath (2009). "A new proof of the density Hales-Jewett theorem". arXiv:0910.3926 [math.CO].

[14] http://michaelnielsen.org/polymath1/index.php?title=Polymath1

[15] http://numberwarrior.wordpress.com/2009/03/25/a-gentle-introduction-to-the-polymath-project/

[16] http://www.tricki.org

[17] Gowers, Timothy (2009-04-16). "Tricki now fully live" (http://gowers.wordpress.com/2009/04/16/tricki-now-fully-live/). . Retrieved 2009-04-16.

[18] Tao, Terence (2009-04-16). "Tricki now live" (http://terrytao.wordpress.com/2009/04/16/tricki-now-live/). What's new. . Retrieved 2009-04-16.

[19] "Status update" (http://gowers.wordpress.com/2010/11/30/status-update/). *Gowers's Weblog*. Timothy Gowers. . Retrieved 1 December 2010.

External links

- Timothy Gowers's blog (http://gowers.wordpress.com/)
- Timothy Gowers's web page (http://www.dpmms.cam.ac.uk/~wtg10/)
- Timothy Gowers (http://genealogy.math.ndsu.nodak.edu/id.php?id=67729) at the Mathematics Genealogy Project.
- Timothy Gowers's results (http://www.imo-official.org/participant_r.aspx?id=11101) at the International Mathematical Olympiad
- 1998 Fields Medalist William Timothy Gowers (http://www.ams.org/featurecolumn/archive/gowers.html) from the American Mathematical Society
- Video lectures by Timothy Gowers on Computational Complexity and Quantum Computation (http://sms.cam.ac.uk/collection/545358)
- Timothy Gowers – Faces of Mathematics (http://www.ma.hw.ac.uk/~ndg/fom/gowersqu.html)
- BBC News (1998): British academics Tim Gowers and Richard Borcherds win top maths awards (http://news.bbc.co.uk/1/hi/education/153535.stm)
- "Multiplying and dividing by whole numbers: Why it is more difficult than you might think" (http://www.gresham.ac.uk/event.asp?PageId=45&EventId=607), lecture by Timothy Gowers at Gresham College, 22 May 2007 (available for download as video and audio files)
- Körner, Tom (1999-09). "Interview with Tim Gowers (Cambridge)" (http://www.emis.de/newsletter/newsletter33.pdf). *Newsletter of the European Mathematical Society* (33): 8–9.
- William Timothy Gowers (http://planetmath.org/?op=getobj&from=objects&id=8367) on PlanetMath
- Listen to Timothy Gowers on The Forum, BBC World Service Radio (http://www.bbc.co.uk/worldservice/indepth/the_forum.shtml)

Jeremy_Gray

Jeremy Gray	
Born	April 25, 1947
Residence	Warwick, UK
Institutions	Open University University of Warwick
Doctoral advisor	Ian Stewart and David Fowler
Doctoral students	June Barrow-Green Snezana Lawrence

Jeremy Gray is an English mathematician primarily interested in the history of mathematics.

He studied mathematics at Oxford University from 1966 to 1969, and then at Warwick University, obtaining his Ph.D. in 1980 under the supervision of Ian Stewart and David Fowler. He has worked at the Open University since 1974, and became a lecturer there in 1978. He also lectures at the University of Warwick since 2000, teaching a course on the history of mathematics.

Gray was a consultant on the television series, *The Story of Maths,*[1] a co-production between the Open University and the BBC.[2]

Books

Gray has been awarded prizes for his contributions to mathematics, including the Albert Leon Whiteman Memorial Prize from the American Mathematical Society.[3]

He has written many books, including:

- *Ideas of space: Euclidean, non-Euclidean, and relativistic*
- *Linear differential equations and group theory from Riemann to Poincaré*
- *The Hilbert Challenge*
- *Janos Bolyai, non-Euclidean Geometry and the Nature of Space*
- *Worlds out of Nothing; a course on the history of geometry in the 19th century*
- *Plato's Ghost: The Modernist Transformation of Mathematics*

He has also contributed to other books:

- *The Princeton Companion to Mathematics*
- *The Oxford Handbook of the History of Mathematics*
- *Revolutions in Mathematics*

Notes

[1] *To Infinity and Beyond* 27 October 2008 21:00 BBC Four
[2] Guardian interview (http://www.guardian.co.uk/science/blog/audio/2008/oct/13/science-weekly-podcast)
[3] http://www2.warwick.ac.uk/fac/sci/maths/general/news/2009/#jgray

External links

- Jeremy Gray (http://genealogy.math.ndsu.nodak.edu/id.php?id=29337) at the Mathematics Genealogy Project.
- Homepage of Jeremy Gray (http://puremaths.open.ac.uk/pmd_department/pmd_gray/index.htm)

Ralph_Greatorex

Ralph Greatorex (c.1625–1675),[1] was a mathematical instrument maker. He was an apprentice of London clockmaker Elias Allen.[1]

Greatorex is mentioned in John Aubrey's Brief Lives as a great friend of William Oughtred the mathematician.[2] He is also briefly referred to in Aubrey's 'Natural History of Wilts', and in the 'Macclesfield Correspondence'.[2] John Evelyn met Greatorex on 8 May 1656, and saw his 'excellent invention to quench fire.'[2] His name appears in 'Pepys's Diary.'[2] On 11 October 1660, when several engines were shown at work in St. James's Park, 'above all the rest,' says Samuel Pepys, 'I liked that which Mr. Greatorex brought, which do carry up the water with a great deal of ease.'[2]

On 24 October Pepys bought of Greatorex a drawing-pen, 'and he did show me the manner of the lamp-glasses which carry the light a great way, good to read in bed by, and I intend to have one of them. And we looked at his wooden jack in his chimney, that goes with the smoake, which indeed is very pretty.'[2] On 9 June and 20 September 1662 and 23 March 1663 ('this day Greatorex brought me a very pretty weather-glasse for heat and cold') Pepys met the inventor; the last entry, 23 May 1663, refers to his varnish, 'which appears every whit as good upon a stick which he hath done, as the Indian.'[2]

As well as Pepys, Allen, Oughtred and Evelyn, others that Greatorex worked or corresponded with included Samuel Hartlib, Christopher Wren, Robert Boyle, Edward Phillips and Jonas Moore.[1] He also attended meetings of the Royal Society, and did horticultural experiments at Arundel House.[1]

References

[1] Sarah Bendall, 'Greatorex, Ralph (c.1625–1675)', Oxford Dictionary of National Biography, Oxford University Press, 2004; online edn, Jan 2008 accessed 6 Feb 2011 (http://www.oxforddnb.com/view/article/11365,)
[2] Greatorex, Ralph, Lydia Miller Middleton, Dictionary of National Biography, 1885-1900, Volume 23

"Greatorex, Ralph". *Dictionary of National Biography*. London: Smith, Elder & Co. 1885–1900.

John_Greaves

John Greaves (1602 – 8 October 1652) was an English mathematician, astronomer and antiquary.

Life

He was born in Colemore, near Alresford, Hampshire. He was the eldest son of John Greaves, rector of Colemore, and Sarah Greaves.[1] His brothers were Nicholas Greaves, Thomas Greaves and Sir Edward Greaves, physician to Charles II.

His father ran a school for sons of the neighbouring gentry, where Greaves began his education.[2] Aged 15, he went to Balliol College, Oxford between 1617–1621, gaining a B.A. degree. In 1624 he was the first of five newly-elected Fellows of Merton College, becoming M.A. in 1628.[3] He began to study astronomy and oriental languages, and especially the works of the ancient eastern astronomers. In 1630 Greaves was chosen professor of geometry at Gresham College, London. Through his predecessor, Peter Turner, he later met archbishop William Laud, the chancellor of Oxford University and Visitor (patron) of Merton College. Laud was keen to make English editions of Greek and Arabic authors, and Greaves' later travels abroad involved collecting manuscripts and books for presentation to his new patron.

Greaves enrolled at University of Leyden in 1633, where he became friends with Jacob Golius, professor of Arabic at Leyden. He enrolled at the University of Padua in 1635 along with George Ent, meeting the Dane Johan Rode (John Rhodius), an expert on ancient weights and measures, who also made a commentary on Celsus. A brief return to England was followed by a second European journey; in 1636 he sailed via Livorno (Leghorn) to Rome, dining with Ent on 5 October at the English College, Rome; he also met William Harvey, who was entertained at the College on the 12th, Gasparo Berti, Lucas Holstenius and Athanasius Kircher. Probably in the same month he met and consulted with the Earl of Arundel's art-collecting agent, William Petty (who dined at the College on 14 October), on the Earl's attempted acquisition of the Obelisk of Domitian, then still lying broken in the Circus of Maxentius. Though 'now it is broken into 5 stones' he measured these and including a sketch of the obelisk as hypothetically repaired in his almanac-notebook (Bodleian Library Savile MS 49,1). Though Arundel paid a 60 crown deposit for the obelisk, pope Urban VIII vetoed its export and it was erected by his successor Innocent X above Bernini's *Fountain of the Four Rivers* in the Piazza Navona.[4]

Greaves toured the Catacombs and made drawings of the Pantheon and Pyramid of Cestius. During his stay in Rome he instituted inquiries into the ancient weights and measures that are among the early classics of metrology.

In 1637 he made a journey to the Levant, one (unfulfilled) intention being to fix the latitude of Alexandria where Ptolemy had made his astronomical observations. He sailed from England to Livorno in the company of Edward Pococke; after a brief visit to Rome, he arrived in Istanbul (Constantinople) around April 1638. There he made the acquaintance of the English ambassador Sir Peter Wyche. He procured various manuscripts there, including a copy of Ptolemy's Almagest ("the fairest work I ever saw").[5] Greaves ended up owning two copies of the Almagest. He was going to have visited the many monastic libraries at Mount Athos, in order to make a catalogue of their MSS and unprinted books. Athos was normally open only to members of the Orthodox church, but thanks to a special dispensation from the Patriarch of Constantinople Cyril Lucaris, Greaves would have had access; but the execution of the patriarch by strangulation in June 1638 for treason against Sultan Murad IV prevented his journey.[3]

Instead, Greaves continued on to Alexandria, where he collected a number of Arabic, Persian and Greek manuscripts. He was an inveterate note-taker, making countless observations in notebooks and on blank pages of other books he bought; he also visited Cairo twice, and made a more accurate survey of the pyramids of Egypt than any traveller who had preceded him. He returned to England in 1640.

On the death of John Bainbridge in 1643, Greaves was appointed as Savilian professor of astronomy and senior reader of the Linacre lecture at Oxford; but he was deprived of his Gresham professorship for having neglected its duties. In 1645 he essayed a reformation of the calendar; but although his plan (of omitting the bissextile day for the next 40 years) was approved by the king, the matter was dropped owing to the turbulent times.

Ejection from Oxford

In 1642 Greaves had been elected subwarden of Merton College. Merton was the only Oxford college to side with the Parliamentarians in the English Civil War, through an earlier dispute in 1638 between Nathaniel Brent, the Warden of Merton, and Greaves' patron William Laud. Brent had been a hostile witness at Laud's 1644 trial. After Laud was executed on 10 January 1645, Greaves drew up a petition for Brent's removal from office; Brent was deposed by Charles I on 27 January.

However, in 1647 a parliamentary commission (visitation) was set up by Parliament "for the correction of offences, abuses, and disorders" in the University of Oxford. Nathaniel Brent was the president of the visitors.[6] After Thomas Fairfax had captured Oxford for the Parliamentarians in 1648 and Brent had returned from London, Greaves was accused of sequestrating the college's plate and funds for king Charles.[7] Despite a deposition from his brother Thomas, Greaves had lost both his Merton fellowship and his Savilian chair by 9 November 1648. Many of his books and MSS disappeared after his rooms were rifled by soldiers, although his friend John Selden managed to recover some of them. However, Greaves was not actually deprived of the professorship until August 1649.[8] He was succeeded as Savilian Professor of Astronomy in that year by Seth Ward, who ensured that Greaves was paid the arrears (£500) of his salary; Greaves was unlikely to have got his money, since the Savilian professors were paid from the income of lands held in Kent and Essex, which were under control of Parliament (rather than the king). Ward also gave over a considerable amount of his own salary to Greaves.[9]

But Greaves' private fortune more than sufficed for all his wants till his death; he retired to London, married and occupied his leisure writing and editing books and manuscripts. He died in London aged 50, and was buried in the church of St Benet Sherehog, which was destroyed during the Great Fire of London.[7]

His brother Nicholas Greaves was his executor.[10] He left his cabinet containing his coin collection to Sir John Marsham, and his astronomical instruments to the Savilian library, now part of the Bodleian Library.[11] Two of his astrolabes (inscribed by his brother Nicholas) are in the Museum of the History of Science, Oxford. (see External links)

Works

Besides his papers in the *Philosophical Transactions*, the principal works of Greaves are:

- *Pyramidographia, or a Description of the Pyramids in Ægypt* (1646)
- *A Discourse on the Roman Foot and Denarius* (1647)
- *Elementa Linguae Persicae* (1649)

The first two were reprinted (together with a biographical notice of the author) as part of *Miscellaneous works of Mr. John Greaves.*[12] This also contains a reprint of

- Withers, Robert (ed. John Greaves) *A Description of the Grand Signour's Seraglio; or Turkish Emperours Court* (1653) London: Printed for Jo. Ridley, at the Castle in Fleet-street by Ram-Alley.

Translations:

- *Chorasmiae et Mawaralnahrae, etc.* (1650). A description of Khwarezm and Mawarannahr (Transoxiana) translated from the Arabic of Abu'l-Fida, with tables of latitude and longitude for the principal towns.
- *Abulfedae Peninsulam Arabum.* This is an edited translation of part of Abu'l-Fida's *History.*
- *Binae Tabulae Geographicae*, two tables of latitude and longitude translated from the Persian of Nasir al-Din al-Tusi and Ulugh Beg

These three works appeared in Hudson, John (1712): Geographiae Veteris Scriptores Graeci Minores [13] Vol. III, Oxon. (in Latin)

- *Lemmata Archimedis, apud Graecos et Latinos iam pridem desiderata, e vetusto codice manuscripto Arabico...* This is a translation of part of Nasir al-Din al-Tusi's edition of the so-called *Middle Books.*[14] Greaves' unpublished MS was revised and amended by Samuel Foster and appeared posthumously in 1659.[15]

The following work is probably not by John Greaves:

- *The origin and antiquity of our English weights and measures discover'd...* [16] (1745 edition) London: Printed for G. Sawbridge. This book, printed in 1706 and again in 1727 and 1745, discusses Greaves' findings and measurements, but is probably the work of someone else.

See also

- List of Gresham Professors of Geometry
- Pyramid inch

References

[1] Shalev, Zur (2005). *The travel notebooks of John Greaves* (http://research.haifa.ac.il/~zshalev/Articles/Shalev-GreavesNotebooks.pdf) in Intersections: Yearbook for early modern studies Vol. 5, 2005. Leiden: Koninklijke Brill NV.

[2] Macquorn Rankine, W. J. in *Imperial Dictionary of Universal Biography* (http://www.archive.org/stream/imperialdictiona02eadi#page/712/mode/2up) Vol II, p.713 London: William Mackenzie

[3] Birch, Thomas (1737). *Miscellaneous works of Mr. John Greaves* (http://www.archive.org/details/miscellaneouswo13unkngoog). London: J. Brindley and C. Corbett. p. i-lxxii. . Retrieved 6 May 2011.

[4] Edward Chaney, "Roma Britannica and the Cultural Memory of Egypt: Lord Arundel and the Obelisk of Domitian", in *Roma Britannica: Art Patronage and Cultural Exchange in Eighteenth-Century Rome*, eds. D. Marshall, K. Wolfe and S. Russell, British School at Rome, 2011, pp. 147–70, fig. 11.11.

[5] Greaves, John (1647); *A Discourse on the Roman Foot and Denarius* London: Wm Lee. p.47

[6] *Dictionary of National Biography*, article on Brent, Sir Nathaniel, pp. 262-4

[7] Ward, John (1740). *The Lives of the Professors of Gresham College, to which is prefixed the Life of the Founder, Sir Thomas Gresham* (http://books.google.co.uk/books?id=jp5bAAAAQAAJ), pp. 144-146 London: John Moore. Google Books full view, retrieved 10 May 2011

[8] Twells, Leonard (1816). *The Lives of Dr. Edward Pocock: the celebrated orientalist, Volume 1* (http://books.google.co.uk/books?id=FvI7AAAAYAAJ). London: Printed for F.C. and J. Rivington, by R. and R. Gilbert. p. 123. . Retrieved 13 May 2011.

[9] Pope, Walter (1697). *The Life of Seth Ward, Lord Bishop of Salisbury* (http://books.google.co.uk/books?id=aU46AAAAcAAJ&PA=18). London: Wm. Keeblewhite. pp. 18–21. . Retrieved 11 May 2011.

[10] *Abstracts of probate acts in the Prerogative court of Canterbury* (Volume 6) (http://www.ebooksread.com/authors-eng/church-of-england-province-of-canterbury-preroga/abstracts-of-probate-acts-in-the-prerogative-court-of-canterbury-volume-6-ruh/page-9-abstracts-of-probate-acts-in-the-prerogative-court-of-canterbury-volume-6-ruh.shtml). Will proved 19th? October 1652.

[11] Smith, Thomas (1707). *Vita quorundam eruditissimorum virorum* (http://books.google.co.uk/books?id=PwIVAAAAQAAJ) (in Latin), p.34 London: David Mortier, at the sign of Erasmus. Google Books full view, retrieved 10 May 2011

[12] Greaves, John (1737) Miscellaneous works of Mr. John Greaves Vol. I (http://books.google.com/books/?id=Puk0AAAAMAAJ) and Vol. II (http://books.google.com/books?id=0uk0AAAAMAAJ). London: Published by Dr Thomas Birch, printed by J. Hughes for J. Brindley and C. Corbett.

[13] http://books.google.com/books?id=fq0-AAAAcAAJ

[14] Hogendijk, Jan P. (2008). "Two Beautiful Geometrical Theorems by Abū Sahl Kūhī in a 17th Century Dutch Translation" (http://www.xs4all.nl/~nirmala/Tarikhelm1.pdf) (PDF). *Tārīkh-e 'Elm: Iranian Journal for the History of Science* (History of Science Institute of the University of Tehran) **6**: 1–36. . Retrieved 8 August 2011.

[15] Foster, Samuel (1659). *Miscellanea sive lucubrationes mathematicae*. London: printed by R. & W. Leybourn, published by John Twysden

[16] http://books.google.co.uk/books?id=0jhbAAAAQAAJ

- This article incorporates text from a publication now in the public domain: Chisholm, Hugh, ed (1911). *Encyclopædia Britannica* (11th ed.). Cambridge University Press.

External links

- Measurer of all things: John Greaves – The great Pyramid and early Modern Metrology by [[Zur Shalev (http://muse.jhu.edu/login?uri=/journals/journal_of_the_history_of_ideas/v063/63.4shalev.pdf)]]
- Astrolabe belonging to John Greaves (http://www.mhs.ox.ac.uk/epact/catalogue.php?ENumber=94032), inscribed by Nicholas Greaves.
- Large astrolabe originally made for Queen Elizabeth I (http://www.mhs.ox.ac.uk/epact/catalogue.php?ENumber=66960), Museum of the History of Science, Oxford

Ben_J._Green

Ben J. Green	

Born	27 February 1977Bristol, England
Residence	Cambridge, England
Nationality	British
Fields	Mathematics
Institutions	University of Bristol University of Cambridge
Alma mater	Trinity College, Cambridge
Doctoral advisor	Timothy Gowers
Notable awards	Clay Research Award (2004) Salem Prize (2005) SASTRA Ramanujan Prize (2007) EMS Prize (2008) Fellow of the Royal Society (2010)

Ben Joseph Green FRS (born 27 February 1977, Bristol, England) is a British mathematician, specializing in combinatorics and number theory. He is the Herchel Smith Professor of Pure Mathematics at the University of Cambridge.

Early years

Ben Green was born on 27 February 1977 in Bristol, England. He studied at local schools of Bristol, Bishop Road Primary School and Fairfield Grammar School, competing in the International Mathematical Olympiad in 1994 and 1995.[1] He entered Trinity College, University of Cambridge in 1995 and completed his B.A. in mathematics in 1998. He earned his doctorate under English mathematician Timothy Gowers in 2003, with a thesis entitled *Topics in arithmetic combinatorics*. He was a research Fellow at Trinity College, Cambridge between 2001 and 2005, before becoming a Professor of Mathematics at the University of Bristol from January 2005 to September 2006. In September 2006 he returned to Cambridge as the first Herchel Smith Professor of Pure Mathematics. He was also a Research Fellow of the Clay Mathematics Institute and held various positions at institutes such as Princeton University, University of British Columbia, and Massachusetts Institute of Technology.

Mathematics

Green has published several important results in both combinatorics and number theory. These include improving the estimate by Jean Bourgain of the size of arithmetic progressions in sumsets, as well as a proof of the Cameron–Erdős conjecture on sum-free sets of natural numbers.

His work in demonstrating that every set of primes of positive relative upper density contains an arithmetic progression of length three then led to his breakthrough 2004 work with mathematician Terence Tao now known as the Green–Tao theorem. This theorem showed that for all n there exist infinitely many arithmetic progressions of length n in the prime numbers.

Awards and honors

Green received the Clay Research Award in 2004 and the Salem Prize in 2005 for his contributions to combinatorial number theory related to progressions of primes.

In 2007 he was awarded the SASTRA Ramanujan Prize.

In 2008 he was among the ten recipients of the prestigious EMS prize awarded every four years only to highly talented young mathematicians who work in Europe.

In 2010 he was elected a Fellow of the Royal Society, who say "Ben Green is a rising star, and universally acknowledged as one of the best young mathematicians in the world."[2]

Notes

[1] Ben J. Green's results (http://www.imo-official.org/participant_r.aspx?id=1137) at the International Mathematical Olympiad

[2] bio-data at (http://royalsociety.org/people/ben-green/) the Royal Society

External links

- Ben Green Homepage (http://www.dpmms.cam.ac.uk/~bjg23/)
- Clay Research Award announcement (http://www.claymath.org/research_award/Green/)
- math.NT/0404188 - Preprint on arbitrarily long arithmetic progressions on primes (http://front.math.ucdavis.edu/math.NT/0404188)
- Ben J. Green (http://genealogy.math.ndsu.nodak.edu/id.php?id=76979) at the Mathematics Genealogy Project.

George_Green

George Green	
Born	14 July 1793Sneinton, Nottinghamshire, England
Died	31 May 1841 (aged 47)Nottingham, Nottinghamshire, England
Alma mater	University of Cambridge
Known for	Green's Theorem Green's Functions Green's Identities Green measure Green's matrix
Influences	John Toplis (speculative)
Influenced	Lord Kelvin Julian Schwinger

George Green (14 July 1793 – 31 May 1841) was a British mathematical physicist who wrote *An Essay on the Application of Mathematical Analysis to the Theories of Electricity and Magnetism* (Green, 1828).[1] [2] The essay introduced several important concepts, among them a theorem similar to the modern Green's theorem, the idea of potential functions as currently used in physics, and the concept of what are now called Green's functions. George Green was the first person to create a mathematical theory of electricity and magnetism and his theory formed the foundation for the work of other scientists such as James Clerk Maxwell, William Thomson, and others. His work ran parallel to that of the great mathematician Gauss (potential theory).

Green's life story is remarkable in that he was almost entirely self-taught. He was born and lived for most of his life in the English town of Sneinton, Nottinghamshire, nowadays part of the city of Nottingham. His father (also named George) was a baker who had built and owned a brick windmill used to grind grain. The younger Green only had about one year of formal schooling as a child, between the ages of 8 and 9.

Early life

In his youth, George Green was described as having a frail constitution and a dislike for doing work in his father's bakery. He had no choice in the matter, however, and as was common for the time he likely began working daily to earn his living at the age of five.

Robert Goodacre's Academy

Roughly 25-50% of children in Nottingham received any schooling in this period. The majority of schools were Sunday schools, run by the Church, and children would typically attend for one or two years only. Recognizing the young Green's above average intellect, and being in a strong financial situation due to his successful bakery, his father enrolled him in March 1801 at Robert Goodacre's Academy in Upper Parliament Street. Robert Goodacre was a well-known science popularizer and educator of the time. He published *Essay on the Education of Youth*, in which he wrote that he did not "study the interest of the boy but the embryo Man". To a non-specialist, he would have seemed deeply knowledgeable in science and maths, but a close inspection of his essay and curriculum revealed that the extent of his mathematical teachings was limited to algebra, trigonometry and logarithms. Thus, George Green's later mathematical contributions, which exhibited knowledge of very modern developments in maths, could not have resulted from his tenure at the Robert Goodacre Academy. He stayed for only four terms (one year), and it was speculated by his contemporaries that he probably exhausted all they had to teach him.

Move from Nottingham to Sneinton

At the time when George's father moved there, 1773, Nottingham had a reputation for being a pleasant town with open spaces and wide roads. By 1831, however, the population had increased nearly five times (in part due to the budding industrial revolution), and Nottingham became known as one of the worst slums in England. There were frequent riots by starving workers, often associated with special hostility towards bakers and millers on the suspicion that they were hiding grain to drive up food prices.

For these reasons, in 1807, George Green senior bought a plot of land in Sneinton, a small town about a mile away from Nottingham. On this plot of land he built a "brick wind corn mill", the wind-mill now famously referred to as Green's Windmill. It was technologically impressive for its time, but required nearly twenty-four hour maintenance, which was to become George Green's burden for the next twenty years.

Green's Windmill

Adult life

Life as a miller

Just as with baking, Green found the responsibilities of operating the mill annoying and tedious. Grain from the fields was arriving continuously at the mill's doorstep, and one had to constantly adjust the sails of the windmill to the windspeed (either to prevent damage, in high winds, or to maximize rotational speed in low winds). The mill-stones that would continuously grind against each other, could wear down or cause a fire (due to friction) if they ran out of grain to grind. Every month the stones, which weighed over a ton, would have to be replaced or repaired.

Family Life

In 1823 George Green formed a relationship with Jane Smith, the daughter of William Smith, hired by Green Senior as mill manager. Although George Green and Jane Smith never married, Jane eventually became known as Jane Green and the couple had seven children together; all but the first had Green as a baptismal name. The final child was born 13 months before Green's death. George Green provided for his common-law wife and children in his will. [3]

Membership to the Nottingham Subscription Library

When Green was thirty, he became a member to the Nottingham Subscription Library. This library exists today, and was likely one of the only sources of Green's advanced mathematical knowledge. Unlike more conventional libraries, the subscription library was exclusive to a hundred or so subscribers (i.e., the first on the list of subscribers was the Duke of Newcastle). This library catered to requests for specialized books and journals that satisfied the particular interests of their subscribers.

The 1828 Essay

In 1828, Green published *An Essay on the Application of Mathematical Analysis to the Theories of Electricity and Magnetism*, which is the essay he is most famous for today. When Green published his *Essay*, it was sold on a subscription basis to 51 people, most of whom were friends and probably could not understand it. The wealthy landowner and mathematician Edward Bromhead bought a copy and encouraged Green to do further work in mathematics. Not believing the offer was sincere, Green did not contact Bromhead for two years.

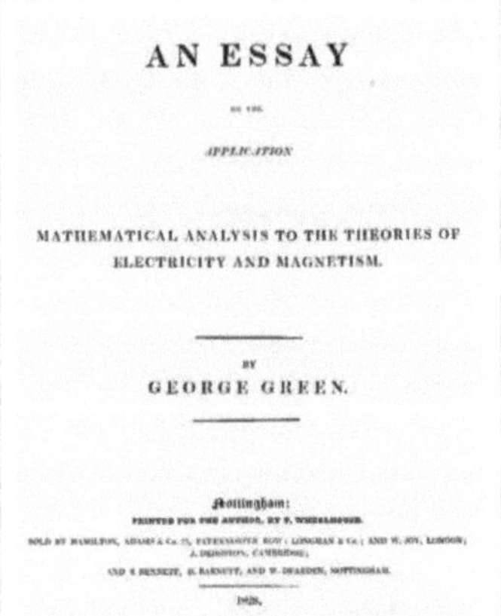
AN ESSAY

ON THE

APPLICATION

OF

MATHEMATICAL ANALYSIS TO THE THEORIES OF ELECTRICITY AND MAGNETISM.

BY

GEORGE GREEN.

Nottingham:

The title page to George Green's original essay on what is now known as Green's theorem. It was published privately at the author's expense, because he thought it would be presumptuous for a person like himself, with no formal education in mathematics, to submit the paper to an established journal.

From miller to mathematician

By 1829, the time when Green's father died, the senior Green had become one of the gentry due to his considerable accumulated wealth and land owned, roughly half of which he left to his son and the other half to his daughter. The young Green, now thirty-six years old, consequently was able to use this wealth to abandon his miller duties and pursue mathematical studies.

Undergraduate at Cambridge

Members of the Nottingham Subscription Library who knew Green repeatedly insisted that he obtain a proper University education. In particular, the most prestigious subscriber (after the Duke of Newcastle) was Sir Edward Bromhead, with whom Green shared many correspondences; he insisted that Green go to Cambridge.

So, in 1832, aged nearly forty, Green was admitted as an undergraduate at Gonville and Caius College, Cambridge.[4] He was particularly insecure about his lack of knowledge of Greek or Latin, which was a prerequisite, but it turned out not to be as hard for him to learn as he had expected (and the expected mastery was not as high as he expected). In the mathematics examinations, he won the first-year mathematical prize. He graduated BA in 1838 as a 4th Wrangler (the 4th highest scoring student in his graduating class, coming after James Joseph Sylvester who scored 2nd).[4]

College fellow

Following his graduation, Green was elected a fellow of the Cambridge Philosophical Society. Even without his stellar academic standing, the Society had already read and made note of his Essay and three other publications, and so Green was warmly welcomed. The next two years provided an unparalleled opportunity for Green to read, write and discuss his scientific ideas. In this short time he published an additional six publications with applications to hydrodynamics, sound and optics.

Final years and posthumous fame

The Grave of George Green, in the church not far from his mill

In his final years at Cambridge, Green became rather ill, and in 1840 he returned to Sneinton, only to die a year later. There are rumours that at Cambridge, Green had "succumbed to alcohol", and some of his earlier supporters, such as Sir Edward Bromhead, tried to distance themselves from him.

Green's work was not well known in the mathematical community during his lifetime. Besides Green himself, the first mathematician to quote his 1828 work was the Briton Robert Murphy (1806–1843) in his 1833 work. In 1845 (four years after Green's death), Green's work was rediscovered by the young William Thomson (age 21 in 1845), later known as Lord Kelvin, who popularised it for future mathematicians. According to the book "George Green" by D.M. Cannell, William Thomson noticed Murphy's citation of Green's 1828 essay but found it difficult to locate Green's 1828 work; he finally got some copies of Green's 1828 work from William Hopkins in 1845. So Green died without any idea of his posthumous fame.

Green's work on the motion of waves in a canal anticipates the WKB approximation of quantum mechanics, while his research on light-waves and the properties of the ether produced what is now known as the Cauchy-Green tensor. Green's theorem and functions were important tools in classical mechanics, and were revised by Schwinger's 1948 work on electrodynamics that led to his 1965 Nobel prize (shared with Feynman and Tomonaga). Green's functions later also proved useful in analyzing superconductivity. On a visit to Nottingham in 1930, Albert Einstein commented that Green had been 20 years ahead of his time. The theoretical physicist, Julian Schwinger, who used Green's functions in his ground-breaking works, published a tribute, entitled "The Greening of Quantum Field Theory: George and I," in 1993.

The George Green Library at the University of Nottingham is named after him, and houses the majority of the University's Science and Engineering Collection. In 1986, Green's Windmill was restored to working order. It now serves both as a working example of a 19th century windmill and as a museum and science centre dedicated to Green.

Westminster Abbey has a memorial stone for George Green in the nave adjoining the graves of Sir Isaac Newton and Lord Kelvin.[5]

List of publications

- *An Essay on the Application of Mathematical Analysis to the Theories of Electricity and Magnetism.* By George Green, Nottingham. Printed for the Author by T. Wheelhouse, Nottingham. 1828. (Quarto, vii + 72 pages.)
- *Mathematical Investigations concerning the Laws of the Equilibrium of Fluids analogous to the Electric Fluid*, with other similar Researches. By George Green, Esq., Communicated by Sir Edward Ffrench Bromhead, Bart., M.A., F.R.S.L. and E. (Cambridge Philosophical Society, read 12 November 1832, printed in the Transactions 1833. Quatro, 64 pages.) Vol. III, Part I.
- *On the Determination of the Exterior and Interior Attractions of Ellipsoids of Variable Densities.* By George Green, Esq., Caius College. (Cambridge Philosophical Society, read 6 May 1833, printed in the Transactions 1835. Quarto, 35 pages.) Vol. III, Part III.
- *Researches on the Vibration of Pendulums in Fluid Media.* By George Green, Esq., Communicated by Sir Edward Ffrench Bromhead, Bart., M.A., F.R.S.S. Lond. and Ed. (Royal Society of Edinburgh, read 16 December 1833, printed in the Transactions 1836, Quarto, 9 pages.) Vol. III, Part I.
- *On the Motion of Waves in a Variable Canal of Small Width and Depth.* By George Green, Esq., BA, of Caius College. (Cambridge Philosophical Society, read 15 May 1837, printed in the Transactions 1838. Quarto, 6

pages.) Vol. VI, Part IV.

- *On the Reflexion and Refraction of Sound.* By George Green, Esq., BA, of Caius College, Cambridge. (Cambridge Philosophical Society, read 11 December 1837, printed in the Transactions 1838. Quarto, 11 pages.) Vol. VI, Part III.
- *On the Laws of Relexion and Refraction of Light at the common Surface of two non-crystallized Media.* By George Green, Esq., BA, of Caius College. (Cambridge Philosophical Society, read 11 December 1837, printed in the Transactions 1838. Quarto, 24 pages.) Vol. VII, Part I.
- *Note on the Motion of Waves in Canals.* By George Green, Esq., BA, of Caius College. (Cambridge Philosophical Society, read 18 February 1839, printed in the Transactions 1839. Quarto, 9 pages.) Vol. VII, Part I.
- *Supplement to a Memoir on the Reflexion and Refraction of Light.* By George Green, Esq., BA, of Caius College. (Cambridge Philosophical Society, read 6 May 1839, printed in the Transactions 1839. Quarto, 8 pages.) Vol. VII, Part I.
- *On the Propagation of Light in Crystallized Media. By George Green, BA, Fellow of Caius College.* (Cambridge Philosophical Society, read 20 May 1839, printed in the Transactions 1839. Quarto, 20 pages.) Vol. VII, Part II.

Mystery of his mathematical knowledge

It is unclear to historians exactly where Green obtained information on current developments in mathematics, as Nottingham had little in the way of intellectual resources. What is even more mysterious is that Green had used "the Mathematical Analysis", a form of calculus derived from Leibniz that was virtually unheard of, or even actively discouraged, in England at the time (due to Leibniz being a contemporary of Newton who had his own methods that were thus championed in England). This form of calculus, and the developments of mathematicians such as Laplace, Lacroix and Poisson were not taught even at Cambridge, let alone Nottingham, and yet Green had not only heard of these developments, but also improved upon them.

It is speculated that only one person educated in mathematics, John Toplis, headmaster of Nottingham High School 1806–1819, graduate from Cambridge and an enthusiast of French mathematics, is known to have lived in Nottingham at the time.

See also

- Green's identities
- Green's matrix
- Green measure

Notes

[1] This 1828 essay can be found in *Mathematical papers of the late George Green*, edited by N. M. Ferrers. The website for this is given below.

[2] Cannell, D.M. (1999) *George Green: An Enigmatic Mathematician*, American Mathematical Monthly **106**(2), 136-151.

[3] Cannel, D. M. and Lord, N. J.; Lord, N. J (March 1993). "George Green, mathematician and physicist 1793-1841". The Mathematical Gazette (The Mathematical Gazette, Vol. 77, No. 478) 77 (478): 26–51. doi:10.2307/3619259. JSTOR 3619259.

[4] Green, George (http://venn.lib.cam.ac.uk/cgi-bin/search.pl?sur=&suro=c&fir=&firo=c&cit=&cito=c&c=all&tex=GRN832G&sye=&eye=&col=all&maxcount=50) in Venn, J. & J. A., *Alumni Cantabrigienses*, Cambridge University Press, 10 vols, 1922–1958.

[5] http://www.westminster-abbey.org/our-history/people/george-green

References

- I. Grattan-Guinness, 'Green, George (1793–1841)', Oxford Dictionary of National Biography, Oxford University Press, 2004 accessed 26 May 2009 (http://www.oxforddnb.com.turing.library.northwestern.edu/view/article/11381,)
- D.M. Cannell, "George Green mathematician and physicist 1793-1841", The Athlone Press, London, 1993.
- Robert Murphy, "On the inverse method of definite integrals", Transactions of the Cambridge Philosophical Society, vol. 4 (1833), pp. 353–408. (Note: This was the first quotation of Green's 1828 work by somebody other than Green himself.)
- O'Connor, John J.; Robertson, Edmund F., "George Green" (http://www-history.mcs.st-andrews.ac.uk/Biographies/Green.html), *MacTutor History of Mathematics archive*, University of St Andrews.
- "George Green" (http://www.britishscienceassociation.org/web/regionsandbranches/branchactivityinyourarea/Midlands/GeorgeGreen/index.htm). - An excellent on-line source of George Green information
- Green, George (1828). "An Essay on the Application of Mathematical Analysis to the Theories of Electricity and Magnetism". arXiv:0807.0088 [physics.hist-ph].
- "Mathematical papers of the late George Green ... Ed. by N. M. Ferrers." (http://name.umdl.umich.edu/AAN8197.0001.001).
- Cannel, D. M. and Lord, N. J.; Lord, N. J (March 1993). "George Green, mathematician and physicist 1793-1841". *The Mathematical Gazette* (The Mathematical Gazette, Vol. 77, No. 478) **77** (478): 26–51. doi:10.2307/3619259. JSTOR 3619259.
- Challis, L. and Sheard, F.; Sheard, Fred (December 2003). "The Green of Green Functions". *Physics Today* **56** (12): 41–46. doi:10.1063/1.1650227.
- "Green's Mill and Science Centre" (http://www.greensmill.org.uk/) (Web page). Retrieved 22 November 2005.
- Schwinger, Julian (1993). *The Greening of quantum Field Theory: George and I.* arXiv:hep-ph/9310283.

Olinthus_Gregory

Olinthus Gilbert Gregory (29 January 1774 – 2 February 1841[1]) was an English mathematician, author and editor.

He was born on 29 January 1774 at Yaxley in Huntingdonshire. Having been educated by Richard Weston, a Leicester botanist, in 1793 he published a treatise, *Lessons Astronomical and Philosophical*. Having settled at Cambridge in 1796, Gregory first acted as sub-editor on the *Cambridge Intelligencer*, and then opened a booksellers shop. In 1802 he obtained an appointment as mathematical master at the Royal Military Academy, Woolwich through the influence of Charles Hutton, to whose notice he had been brought by a manuscript on the Use of the Sliding Rule; and when Hutton resigned in 1807 Gregory succeeded him in the professorship. Failing health obliged him to retire in 1838, and he died at Woolwich on 2 February 1841.

Gregory's books included *Hints for the Use of Teachers of Elementary Mathematics* (1840, new edition 1853), and *Mathematics for Practical Men* (1825), which was revised and enlarged by Henry Law in 1848, and again by J. R. Young in 1862. His *Letters on the Evidences of Christianity* (1815) have been several times reprinted, and an abridgment was published by the Religious Tract Society in 1853. He will probably be longest remembered for his *Biography of Robert Hall*, which first appeared in the collected edition of Hall's works, was published separately in 1833, and has since passed through several editions. The minor importance of his *Memoir of John Mason Good* (1828) is due to the narrower fame of the subject.

He was one of the founders of the Royal Astronomical Society. In 1802 he was appointed editor of the *Gentlemen's Diary*, and in 1818 editor of the *Lady's Diary* and superintendent of the almanacs of the Stationers Company. Gregory was a co-editor of *Pantologia*, 1813.

Gregory's son was Charles Hutton Gregory, who later became president of the Institution of Civil Engineers.

Notes

[1] "Alexander Gordon, 'Gregory, Olinthus Gilbert (1774–1841)', rev. Ben Marsden, Oxford Dictionary of National Biography, Oxford University Press, 2004 (subscriber access only)" (http://www.oxforddnb.com/view/article/11469). . Retrieved 2008-05-21.

John_Greig_(mathematician)

John Greig (1759–1819), was an English mathematician.

Greig died at Somers Town, London, on 19 January 1819, aged 60 (Gent. Mag. 1819, i. 184). He taught mathematics and wrote:

- 'The Young Lady's Guide to Arithmetic,' London, 1798; many editions, the last in 1864.
- 'Introduction to the Use of the Globes,' 1805 ; three editions.
- 'A New Introduction to Arithmetic,' London, 1805.
- 'A System of Astronomy on the simple plan of Geography,' London, 1810.
- 'Astrography, or the Heavens displayed,' London, 1810.
- 'The World displayed, or the Characteristic Features of Nature and Art,' London, 1810.

References

This article incorporates text from a publication now in the public domain: "Greig, John". *Dictionary of National Biography*. London: Smith, Elder & Co. 1885–1900.

Article Sources and Contributors

Ronald_Brown_(mathematician) *Source*: http://en.wikipedia.org/w/index.php?title=Ronald_Brown_%28mathematician%29 *Contributors*: Bci2, Charles Matthews, Moonriddengirl, Oracleofottawa, Rjwilmsi, William Avery, Yworo, 1 anonymous edits

George_Gargrave *Source*: http://en.wikipedia.org/w/index.php?title=George_Gargrave *Contributors*: Boleyn, Michael Hardy

William_Gascoigne_(scientist) *Source*: http://en.wikipedia.org/w/index.php?title=William_Gascoigne_%28scientist%29 *Contributors*: Aa42john, Angusmclellan, Ast-hist001, Astrochemist, Byron Vickers, CielProfond, D6, Ephraim SC, Etacar11, Fagairolles 34, FeanorStar7, Gene Nygaard, Heron, Inglis, Jaraalbe, Jevansen, John of Reading, Jonnabuz, Kbthompson, Lord Loxley, Lumbercutter, Mais oui!, Michael Daly, Michael David, Mikeblas, Philip Baird Shearer, Richerman, Rjwilmsi, Tom.k, Trappedinburnley, 9 anonymous edits

Henry_Gellibrand *Source*: http://en.wikipedia.org/w/index.php?title=Henry_Gellibrand *Contributors*: Charles Matthews, Mais oui!, Michael Hardy, Oracleofottawa, RockMagnetist, Woohookitty, 2 anonymous edits

Gerland_(mathematician) *Source*: http://en.wikipedia.org/w/index.php?title=Gerland_%28mathematician%29 *Contributors*: Chris the speller, Korg, Markussep, Tim!, Waacstats

Peter_Giblin *Source*: http://en.wikipedia.org/w/index.php?title=Peter_Giblin *Contributors*: 9258fahsflkh917fas, CactusWriter, Eastlaw, Mufka, NativeForeigner, Sadads, Salix alba, Special-T, The Thing That Should Not Be, Tkynerd, Waacstats, 7 anonymous edits

Richard_D._Gill *Source*: http://en.wikipedia.org/w/index.php?title=Richard_D._Gill *Contributors*: A. Carty, Caponer, G716, Gill110951, Gimmetoo, GirasoleDE, Headbomb, Hippalus, Kiefer.Wolfowitz, Koavf, Mayumashu, Michael Hardy, Nsk92, NuclearWarfare, Otto ter Haar, Rjwilmsi, RogDel, Uncompetence, Walgamanus, Wildhartlivie, 1 anonymous edits

James_Whitbread_Lee_Glaisher *Source*: http://en.wikipedia.org/w/index.php?title=James_Whitbread_Lee_Glaisher *Contributors*: Blotwell, Charles Matthews, D6, Dsp13, Enric Naval, Giftlite, Icairns, John Vandenberg, Linas, Magnus Manske, Mais oui!, Michael Hardy, Occuli, Plucas58, Rich Farmbrough, Robinh, Saga City, Suslindisambiguator, TheParanoidOne, Twang, VivaEmilyDavies, 3 anonymous edits

Alfred_Goldie *Source*: http://en.wikipedia.org/w/index.php?title=Alfred_Goldie *Contributors*: Charvest, Giftlite, Omnipaedista, Rschwieb, Woohookitty, 1 anonymous edits

Benjamin_Gompertz *Source*: http://en.wikipedia.org/w/index.php?title=Benjamin_Gompertz *Contributors*: Academic Challenger, Agricola44, Arnejohs, Avraham, Bearcat, Bender235, Bnynms, BrendanH, Charles Matthews, Chicheley, D6, DESiegel, DavidCBryant, Garion96, Gavrilov, Giftlite, Gompertz, Ian Geoffrey Kennedy, Jaime Saldarriaga, Julo, Mack2, Mais oui!, Marcus2, MathMartin, Michael Hardy, Mpulier, Muchness, Necrothesp, Newport, Oleg Alexandrov, Oracleofottawa, Pdn, Porcher, Rjwilmsi, Smallweed, Tuesdaily, Unused, Uscitizenjason, VivaEmilyDavies, Whouk, Xevious, 16 anonymous edits

I._J._Good *Source*: http://en.wikipedia.org/w/index.php?title=I._J._Good *Contributors*: A. Carty, Agricola44, Avenue, Best name, Billjefferys, Blargh29, Bongomatic, Bradtcordeiro, Brianyoumans, Bryan Derksen, Cal T, Charles Matthews, Chrysippo, D6, David Gerard, Eastfrisian, Erasmussen, G716, Glen, Gwern, Hair Commodore, HamburgerRadio, Icairns, Igorpak, J. W. Love, Jaredwf, Jheald, John, Kenneth M Burke, Kiefer.Wolfowitz, Kinston eagle, Kosmocentric, Logologist, MarkSweep, Matt Crypto, Mhym, Michael Hardy, Michaelwilson, Nietzsche 2, Nihil novi, Orgeat, Piotrus, Portia327, Pshoplifter, Qwfp, Rms125a@hotmail.com, Rpyle731, Runcorn, Spageti, Ssd, Stefanomione, TedColes, TexasAndroid, TheParanoidOne, Tomi, TomyDuby, WWGB, Walter Breitzke, Weregerbil, WhiteDragon, Wile E. Heresiarch, Wizardman, X-Kind, 33 anonymous edits

Reuben_Goodstein *Source*: http://en.wikipedia.org/w/index.php?title=Reuben_Goodstein *Contributors*: Anonymous Dissident, Craigy144, EagleFan, Icairns, Lamro, MathMartin, R.e.s., Rholton, Rhythm, Rich Farmbrough, Rjwilmsi, Sligocki, Tassedethe, The Sage of Stamford, Timrollpickering, Unzerlegbarkeit, VivaEmilyDavies, XJamRastafire, 7 anonymous edits

John_Gough_(natural_philosopher) *Source*: http://en.wikipedia.org/w/index.php?title=John_Gough_%28natural_philosopher%29 *Contributors*: Astrochemist, Dr Greg, John, MessinaRagazza, Peterlewis, QueenAdelaide, Rjwilmsi, RobertG, Ronhjones, Wknight94, 6 anonymous edits

Timothy_Gowers *Source*: http://en.wikipedia.org/w/index.php?title=Timothy_Gowers *Contributors*: 7, AdamSmithee, AgarwalSumeet, Akriasas, Allemandtando, AllenS, Almit39, Arthur Rubin, AxelBoldt, Bender235, Bobfish 341, Busy365, Charles Matthews, Citrus538, Cmapm, D6, David Eppstein, DavidCBryant, Djsasso, Dmn, Dominico, Duncan.Hull, DuncanHill, EileenFlammable, Eubulides, Euyyn, Florian Huber, Gandalf61, Gareth Jones, GcSwRhIc, Giftlite, Grstain, Headbomb, Hell9, Hu12, Icairns, Jackyd101, JamesAM, Jamesfranklingresham, Jaraalbe, Jaredwf, Jheald, Jinhyun park, Jleedev, Jodi.a.schneider, Joseph Myers, Jtwdog, Justin W Smith, Kate, Kiefer.Wolfowitz, LJosil, Lord.lucan, Lupin, Luwilt, Mackensen, Markcollinsx, MathMartin, Michael Hardy, Mietchen, Mrh30, Outis9, Piccadilly, Rich Farmbrough, Rjwilmsi, S11-73-3-33, Scope creep, Ser Amantio di Nicolao, Siva1979, Smythu10, Sorumbatico, Swirlix, Thincat, VanBurenen, Vanish2, VivaEmilyDavies, Warrickball, Woohookitty, XJamRastafire, 67 anonymous edits

Jeremy_Gray *Source*: http://en.wikipedia.org/w/index.php?title=Jeremy_Gray *Contributors*: Amielcool, David Eppstein, GcSwRhIc, GoingBatty, Jezhotwells, Keith D, Michael Hardy, Oracleofottawa, Sam Derbyshire, Solzhenitsyn1, Tkuvho, Woohookitty, Þjóðólfr, 1 anonymous edits

Ralph_Greatorex *Source*: http://en.wikipedia.org/w/index.php?title=Ralph_Greatorex *Contributors*: Boleyn, Carcharoth, David Eppstein, RayAYang, Rjwilmsi

John_Greaves *Source*: http://en.wikipedia.org/w/index.php?title=John_Greaves *Contributors*: Boleyn2, Brucelee, CanisRufus, Charles Matthews, FeanorStar7, Hanay, JackyR, Jim62sch, Martarius, McKay, MinorProphet, Pedro, R'n'B, RogDel, Staxringold, Tuesdaily, Wetman, Woohookitty, 8 anonymous edits

Ben_J._Green *Source*: http://en.wikipedia.org/w/index.php?title=Ben_J._Green *Contributors*: Algebraist, Benbwfc89, Bender235, Bluap, Boleyn2, Can't sleep, clown will eat me, Charles Matthews, Chrisdab, Cohesion, Cutler, D6, Dmr2, Domminico, Dr Zimbu, Drummle, Emijrp, Encephalon, Etale, Ex-User777, Gareth Jones, GiantSnowman, Giftlite, Greenyno4, Grey Shadow, Inflammator, J. Van Meter, J.delanoy, JMS Old Al, Jakash, Jiuguang Wang, Joth, JustAGal, Kakofonous, MONGO, Masterpiece2000, Mets501, Michael Slone, Mrh30, NBeale, NawlinWiki, Ntsimp, OlEnglish, Oleg Alexandrov, Princemackenzie, Profangelo, Qwghlm, RJChapman, Rick Block, Ser Amantio di Nicolao, Smb1001, SwisterTwister, Teorth, VanBurenen, Vanish2, Wahoofive, 77 anonymous edits

George_Green *Source*: http://en.wikipedia.org/w/index.php?title=George_Green *Contributors*: 28bytes, A. di M., Aaron Brenneman, Agricola44, Algebraist, Aliotra, Alma Pater, Arjen Dijksman, Art LaPella, Bevo74, Billlion, CALR, CJBR, Carcharoth, Charles Matthews, Chicheley, Chochopk, Chris the speller, ChrisWilmer, Christian75, Curps, Cwkmail, DESiegel, DShamen, Daemonic Kangaroo, David Newton, Drbreznjev, Dsp13, Dysprosia, Epbr123, Erics2392, Fooblitz, Francs2000, Gaius Cornelius, GeoGreg, Georgegreen6, Giftlite, HannsEwald, Headbomb, Hqb, Icairns, JYOuyang, Japanese Searobin, Jaredwf, Jason Davies, Jeff79, JesseHogan, Johnbibby, Justin W Smith, Kpjas, Magioladitis, Markus Poessel, Mathman10, Nicholas Jackson, Omnipaedista, PAR, Palmiped, Paul Magnussen, Paul Matthews, Peculiarist, Philip Brown, Pit, Poor Yorick, Reddi, Rich Farmbrough, Richie, Rjwilmsi, Rmashhadi, Roleplayer, Rubberstamp, SimonArlott, SpNeo, StewartMH, TBond, Tpbradbury, Tricolours, Welsh, Wereon, Wizardman, Woohookitty, Woz2, Wricardoh, XJamRastafire, 66 anonymous edits

Olinthus_Gregory *Source*: http://en.wikipedia.org/w/index.php?title=Olinthus_Gregory *Contributors*: Apwoolrich, CanisRufus, Carbonix, DShamen, Dumelow, Gdr, Good Olfactory, Herbee, Jackyd101, Jaraalbe, Rnt20, Saga City, SemperBlotto, T. Anthony, Tim!, VivaEmilyDavies, Wmahan

John_Greig_(mathematician) *Source*: http://en.wikipedia.org/w/index.php?title=John_Greig_%28mathematician%29 *Contributors*: Boleyn, Buster40004, Michael Hardy, Ser Amantio di Nicolao

Image Sources, Licenses and Contributors

File:Flag of the United Kingdom.svg *Source*: http://en.wikipedia.org/w/index.php?title=File:Flag_of_the_United_Kingdom.svg *License*: unknown *Contributors*: Anomie, Mifter

File:Wikisource-logo.svg *Source*: http://en.wikipedia.org/w/index.php?title=File:Wikisource-logo.svg *License*: unknown *Contributors*: Nicholas Moreau

Image:Gascoigne's micrometer as drawn by Robert Hooke .JPG *Source*: http://en.wikipedia.org/w/index.php?title=File:Gascoigne's_micrometer_as_drawn_by_Robert_Hooke_.JPG *License*: unknown *Contributors*: Robert Hooke

File:RDG110951.jpg *Source*: http://en.wikipedia.org/w/index.php?title=File:RDG110951.jpg *License*: unknown *Contributors*: Gill110951

File:James Whitbread Lee Glaisher.jpg *Source*: http://en.wikipedia.org/w/index.php?title=File:James_Whitbread_Lee_Glaisher.jpg *License*: unknown *Contributors*: Raymond, Skraemer, 2 anonymous edits

File:wikisource-logo.svg *Source*: http://en.wikipedia.org/w/index.php?title=File:Wikisource-logo.svg *License*: unknown *Contributors*: Nicholas Moreau

File:gompertz.png *Source*: http://en.wikipedia.org/w/index.php?title=File:Gompertz.png *License*: unknown *Contributors*: Julo, PMG

Image:Gough1.GIF *Source*: http://en.wikipedia.org/w/index.php?title=File:Gough1.GIF *License*: unknown *Contributors*: Peterlewis

File:Timothy Gowers Washington 2009.jpg *Source*: http://en.wikipedia.org/w/index.php?title=File:Timothy_Gowers_Washington_2009.jpg *License*: unknown *Contributors*: User:AllenS

File:Ben Green.jpg *Source*: http://en.wikipedia.org/w/index.php?title=File:Ben_Green.jpg *License*: unknown *Contributors*: Renate Schmid, Copyright is with MFO

Image:Green's windmill.jpg *Source*: http://en.wikipedia.org/w/index.php?title=File:Green's_windmill.jpg *License*: unknown *Contributors*: Original uploader was Kev747 at en.wikipedia

Image:GreenEssay.png *Source*: http://en.wikipedia.org/w/index.php?title=File:GreenEssay.png *License*: unknown *Contributors*: Jrockley, Kåre-Olav, PAR

File:George Green's Grave - geograph.org.uk - 2160600.jpg *Source*: http://en.wikipedia.org/w/index.php?title=File:George_Green's_Grave_-_geograph.org.uk_-_2160600.jpg *License*: unknown *Contributors*: Ashley Dace

Image: Olinthus Gregory.jpg *Source*: http://en.wikipedia.org/w/index.php?title=File:Olinthus_Gregory.jpg *License*: unknown *Contributors*: Gdr, PMG

MIX
Papier aus verantwortungsvollen Quellen
Paper from responsible sources
FSC® C105338

Printed by Books on Demand GmbH, Norderstedt / Germany